COMPLIANCE ASSURANCE
FOR THE SAFE TRANSPORT OF
RADIOACTIVE MATERIAL

The following States are Members of the International Atomic Energy Agency:

AFGHANISTAN
ALBANIA
ALGERIA
ANGOLA
ANTIGUA AND BARBUDA
ARGENTINA
ARMENIA
AUSTRALIA
AUSTRIA
AZERBAIJAN
BAHAMAS
BAHRAIN
BANGLADESH
BARBADOS
BELARUS
BELGIUM
BELIZE
BENIN
BOLIVIA, PLURINATIONAL
  STATE OF
BOSNIA AND HERZEGOVINA
BOTSWANA
BRAZIL
BRUNEI DARUSSALAM
BULGARIA
BURKINA FASO
BURUNDI
CAMBODIA
CAMEROON
CANADA
CENTRAL AFRICAN
  REPUBLIC
CHAD
CHILE
CHINA
COLOMBIA
COMOROS
CONGO
COSTA RICA
CÔTE D'IVOIRE
CROATIA
CUBA
CYPRUS
CZECH REPUBLIC
DEMOCRATIC REPUBLIC
  OF THE CONGO
DENMARK
DJIBOUTI
DOMINICA
DOMINICAN REPUBLIC
ECUADOR
EGYPT
EL SALVADOR
ERITREA
ESTONIA
ESWATINI
ETHIOPIA
FIJI
FINLAND
FRANCE
GABON
GEORGIA

GERMANY
GHANA
GREECE
GRENADA
GUATEMALA
GUYANA
HAITI
HOLY SEE
HONDURAS
HUNGARY
ICELAND
INDIA
INDONESIA
IRAN, ISLAMIC REPUBLIC OF
IRAQ
IRELAND
ISRAEL
ITALY
JAMAICA
JAPAN
JORDAN
KAZAKHSTAN
KENYA
KOREA, REPUBLIC OF
KUWAIT
KYRGYZSTAN
LAO PEOPLE'S DEMOCRATIC
  REPUBLIC
LATVIA
LEBANON
LESOTHO
LIBERIA
LIBYA
LIECHTENSTEIN
LITHUANIA
LUXEMBOURG
MADAGASCAR
MALAWI
MALAYSIA
MALI
MALTA
MARSHALL ISLANDS
MAURITANIA
MAURITIUS
MEXICO
MONACO
MONGOLIA
MONTENEGRO
MOROCCO
MOZAMBIQUE
MYANMAR
NAMIBIA
NEPAL
NETHERLANDS
NEW ZEALAND
NICARAGUA
NIGER
NIGERIA
NORTH MACEDONIA
NORWAY
OMAN
PAKISTAN

PALAU
PANAMA
PAPUA NEW GUINEA
PARAGUAY
PERU
PHILIPPINES
POLAND
PORTUGAL
QATAR
REPUBLIC OF MOLDOVA
ROMANIA
RUSSIAN FEDERATION
RWANDA
SAINT KITTS AND NEVIS
SAINT LUCIA
SAINT VINCENT AND
  THE GRENADINES
SAMOA
SAN MARINO
SAUDI ARABIA
SENEGAL
SERBIA
SEYCHELLES
SIERRA LEONE
SINGAPORE
SLOVAKIA
SLOVENIA
SOUTH AFRICA
SPAIN
SRI LANKA
SUDAN
SWEDEN
SWITZERLAND
SYRIAN ARAB REPUBLIC
TAJIKISTAN
THAILAND
TOGO
TONGA
TRINIDAD AND TOBAGO
TUNISIA
TÜRKİYE
TURKMENISTAN
UGANDA
UKRAINE
UNITED ARAB EMIRATES
UNITED KINGDOM OF
  GREAT BRITAIN AND
  NORTHERN IRELAND
UNITED REPUBLIC
  OF TANZANIA
UNITED STATES OF AMERICA
URUGUAY
UZBEKISTAN
VANUATU
VENEZUELA, BOLIVARIAN
  REPUBLIC OF
VIET NAM
YEMEN
ZAMBIA
ZIMBABWE

The Agency's Statute was approved on 23 October 1956 by the Conference on the Statute of the IAEA held at United Nations Headquarters, New York; it entered into force on 29 July 1957. The Headquarters of the Agency are situated in Vienna. Its principal objective is "to accelerate and enlarge the contribution of atomic energy to peace, health and prosperity throughout the world".

IAEA SAFETY STANDARDS SERIES No. SSG-78

COMPLIANCE ASSURANCE FOR THE SAFE TRANSPORT OF RADIOACTIVE MATERIAL

SPECIFIC SAFETY GUIDE

INTERNATIONAL ATOMIC ENERGY AGENCY
VIENNA, 2023

IAEA Library Cataloguing in Publication Data

Names: International Atomic Energy Agency.
Title: Compliance assurance for the safe transport of radioactive material / International Atomic Energy Agency.
Description: Vienna : International Atomic Energy Agency, 2023. | Series: safety standards series, ISSN 1020–525X ; no. SSG-78 | Includes bibliographical references.
Identifiers: IAEAL 22-01540 | ISBN 978–92–0–141922–4 (paperback : alk. paper) | ISBN 978–92–0–142022–0 (pdf) | ISBN 978–92–0–142122–7 (epub)
Subjects: LCSH: Radioactive substances — Transportation — Safety measures. | Radioactive substances — Safety regulations. | Radioactive materials.
Classification: UDC 656.073.436 | STI/PUB/2033

FOREWORD

by Rafael Mariano Grossi
Director General

The IAEA's Statute authorizes it to "establish…standards of safety for protection of health and minimization of danger to life and property". These are standards that the IAEA must apply to its own operations, and that States can apply through their national regulations.

The IAEA started its safety standards programme in 1958 and there have been many developments since. As Director General, I am committed to ensuring that the IAEA maintains and improves upon this integrated, comprehensive and consistent set of up to date, user friendly and fit for purpose safety standards of high quality. Their proper application in the use of nuclear science and technology should offer a high level of protection for people and the environment across the world and provide the confidence necessary to allow for the ongoing use of nuclear technology for the benefit of all.

Safety is a national responsibility underpinned by a number of international conventions. The IAEA safety standards form a basis for these legal instruments and serve as a global reference to help parties meet their obligations. While safety standards are not legally binding on Member States, they are widely applied. They have become an indispensable reference point and a common denominator for the vast majority of Member States that have adopted these standards for use in national regulations to enhance safety in nuclear power generation, research reactors and fuel cycle facilities as well as in nuclear applications in medicine, industry, agriculture and research.

The IAEA safety standards are based on the practical experience of its Member States and produced through international consensus. The involvement of the members of the Safety Standards Committees, the Nuclear Security Guidance Committee and the Commission on Safety Standards is particularly important, and I am grateful to all those who contribute their knowledge and expertise to this endeavour.

The IAEA also uses these safety standards when it assists Member States through its review missions and advisory services. This helps Member States in the application of the standards and enables valuable experience and insight to be shared. Feedback from these missions and services, and lessons identified from events and experience in the use and application of the safety standards, are taken into account during their periodic revision.

I believe the IAEA safety standards and their application make an invaluable contribution to ensuring a high level of safety in the use of nuclear technology. I encourage all Member States to promote and apply these standards, and to work with the IAEA to uphold their quality now and in the future.

THE IAEA SAFETY STANDARDS

BACKGROUND

Radioactivity is a natural phenomenon and natural sources of radiation are features of the environment. Radiation and radioactive substances have many beneficial applications, ranging from power generation to uses in medicine, industry and agriculture. The radiation risks to workers and the public and to the environment that may arise from these applications have to be assessed and, if necessary, controlled.

Activities such as the medical uses of radiation, the operation of nuclear installations, the production, transport and use of radioactive material, and the management of radioactive waste must therefore be subject to standards of safety.

Regulating safety is a national responsibility. However, radiation risks may transcend national borders, and international cooperation serves to promote and enhance safety globally by exchanging experience and by improving capabilities to control hazards, to prevent accidents, to respond to emergencies and to mitigate any harmful consequences.

States have an obligation of diligence and duty of care, and are expected to fulfil their national and international undertakings and obligations.

International safety standards provide support for States in meeting their obligations under general principles of international law, such as those relating to environmental protection. International safety standards also promote and assure confidence in safety and facilitate international commerce and trade.

A global nuclear safety regime is in place and is being continuously improved. IAEA safety standards, which support the implementation of binding international instruments and national safety infrastructures, are a cornerstone of this global regime. The IAEA safety standards constitute a useful tool for contracting parties to assess their performance under these international conventions.

THE IAEA SAFETY STANDARDS

The status of the IAEA safety standards derives from the IAEA's Statute, which authorizes the IAEA to establish or adopt, in consultation and, where appropriate, in collaboration with the competent organs of the United Nations and with the specialized agencies concerned, standards of safety for protection of health and minimization of danger to life and property, and to provide for their application.

With a view to ensuring the protection of people and the environment from harmful effects of ionizing radiation, the IAEA safety standards establish fundamental safety principles, requirements and measures to control the radiation exposure of people and the release of radioactive material to the environment, to restrict the likelihood of events that might lead to a loss of control over a nuclear reactor core, nuclear chain reaction, radioactive source or any other source of radiation, and to mitigate the consequences of such events if they were to occur. The standards apply to facilities and activities that give rise to radiation risks, including nuclear installations, the use of radiation and radioactive sources, the transport of radioactive material and the management of radioactive waste.

Safety measures and security measures[1] have in common the aim of protecting human life and health and the environment. Safety measures and security measures must be designed and implemented in an integrated manner so that security measures do not compromise safety and safety measures do not compromise security.

The IAEA safety standards reflect an international consensus on what constitutes a high level of safety for protecting people and the environment from harmful effects of ionizing radiation. They are issued in the IAEA Safety Standards Series, which has three categories (see Fig. 1).

Safety Fundamentals

Safety Fundamentals present the fundamental safety objective and principles of protection and safety, and provide the basis for the safety requirements.

Safety Requirements

An integrated and consistent set of Safety Requirements establishes the requirements that must be met to ensure the protection of people and the environment, both now and in the future. The requirements are governed by the objective and principles of the Safety Fundamentals. If the requirements are not met, measures must be taken to reach or restore the required level of safety. The format and style of the requirements facilitate their use for the establishment, in a harmonized manner, of a national regulatory framework. Requirements, including numbered 'overarching' requirements, are expressed as 'shall' statements. Many requirements are not addressed to a specific party, the implication being that the appropriate parties are responsible for fulfilling them.

Safety Guides

Safety Guides provide recommendations and guidance on how to comply with the safety requirements, indicating an international consensus that it

---

[1]   See also publications issued in the IAEA Nuclear Security Series.

FIG. 1. The long term structure of the IAEA Safety Standards Series.

is necessary to take the measures recommended (or equivalent alternative measures). The Safety Guides present international good practices, and increasingly they reflect best practices, to help users striving to achieve high levels of safety. The recommendations provided in Safety Guides are expressed as 'should' statements.

APPLICATION OF THE IAEA SAFETY STANDARDS

The principal users of safety standards in IAEA Member States are regulatory bodies and other relevant national authorities. The IAEA safety standards are also used by co-sponsoring organizations and by many organizations that design, construct and operate nuclear facilities, as well as organizations involved in the use of radiation and radioactive sources.

The IAEA safety standards are applicable, as relevant, throughout the entire lifetime of all facilities and activities — existing and new — utilized for peaceful purposes and to protective actions to reduce existing radiation risks. They can be used by States as a reference for their national regulations in respect of facilities and activities.

The IAEA's Statute makes the safety standards binding on the IAEA in relation to its own operations and also on States in relation to IAEA assisted operations.

The IAEA safety standards also form the basis for the IAEA's safety review services, and they are used by the IAEA in support of competence building, including the development of educational curricula and training courses.

International conventions contain requirements similar to those in the IAEA safety standards and make them binding on contracting parties. The IAEA safety standards, supplemented by international conventions, industry standards and detailed national requirements, establish a consistent basis for protecting people and the environment. There will also be some special aspects of safety that need to be assessed at the national level. For example, many of the IAEA safety standards, in particular those addressing aspects of safety in planning or design, are intended to apply primarily to new facilities and activities. The requirements established in the IAEA safety standards might not be fully met at some existing facilities that were built to earlier standards. The way in which IAEA safety standards are to be applied to such facilities is a decision for individual States.

The scientific considerations underlying the IAEA safety standards provide an objective basis for decisions concerning safety; however, decision makers must also make informed judgements and must determine how best to balance the benefits of an action or an activity against the associated radiation risks and any other detrimental impacts to which it gives rise.

DEVELOPMENT PROCESS FOR THE IAEA SAFETY STANDARDS

The preparation and review of the safety standards involves the IAEA Secretariat and five Safety Standards Committees, for emergency preparedness and response (EPReSC) (as of 2016), nuclear safety (NUSSC), radiation safety (RASSC), the safety of radioactive waste (WASSC) and the safe transport of radioactive material (TRANSSC), and a Commission on Safety Standards (CSS) which oversees the IAEA safety standards programme (see Fig. 2).

All IAEA Member States may nominate experts for the Safety Standards Committees and may provide comments on draft standards. The membership of the Commission on Safety Standards is appointed by the Director General and includes senior governmental officials having responsibility for establishing national standards.

A management system has been established for the processes of planning, developing, reviewing, revising and establishing the IAEA safety standards.

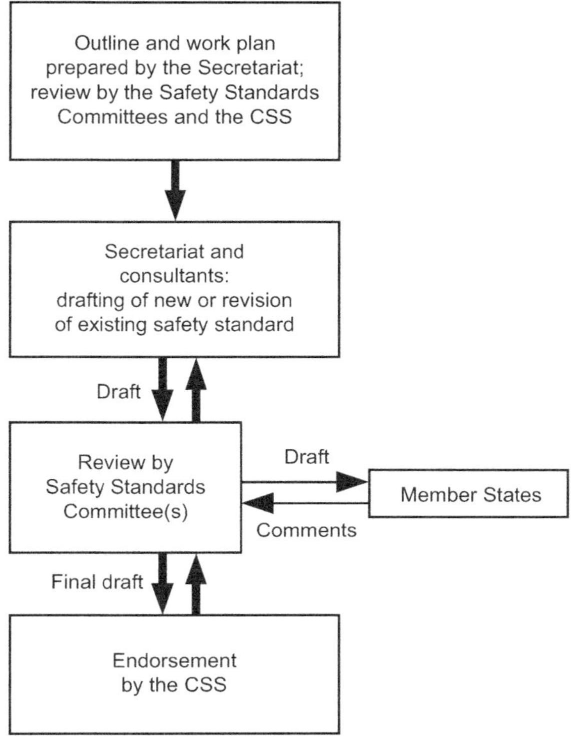

FIG. 2. The process for developing a new safety standard or revising an existing standard.

It articulates the mandate of the IAEA, the vision for the future application of the safety standards, policies and strategies, and corresponding functions and responsibilities.

INTERACTION WITH OTHER INTERNATIONAL ORGANIZATIONS

The findings of the United Nations Scientific Committee on the Effects of Atomic Radiation (UNSCEAR) and the recommendations of international expert bodies, notably the International Commission on Radiological Protection (ICRP), are taken into account in developing the IAEA safety standards. Some safety standards are developed in cooperation with other bodies in the United Nations system or other specialized agencies, including the Food and Agriculture Organization of the United Nations, the United Nations Environment Programme, the International Labour Organization, the OECD Nuclear Energy Agency, the Pan American Health Organization and the World Health Organization.

INTERPRETATION OF THE TEXT

Safety related terms are to be understood as defined in the IAEA Nuclear Safety and Security Glossary (see https://www.iaea. org/resources/publications/iaea-nuclear-safety-and-security-glossary). Otherwise, words are used with the spellings and meanings assigned to them in the latest edition of The Concise Oxford Dictionary. For Safety Guides, the English version of the text is the authoritative version.

The background and context of each standard in the IAEA Safety Standards Series and its objective, scope and structure are explained in Section 1, Introduction, of each publication.

Material for which there is no appropriate place in the body text (e.g. material that is subsidiary to or separate from the body text, is included in support of statements in the body text, or describes methods of calculation, procedures or limits and conditions) may be presented in appendices or annexes.

An appendix, if included, is considered to form an integral part of the safety standard. Material in an appendix has the same status as the body text, and the IAEA assumes authorship of it. Annexes and footnotes to the main text, if included, are used to provide practical examples or additional information or explanation. Annexes and footnotes are not integral parts of the main text. Annex material published by the IAEA is not necessarily issued under its authorship; material under other authorship may be presented in annexes to the safety standards. Extraneous material presented in annexes is excerpted and adapted as necessary to be generally useful.

CONTENTS

1. INTRODUCTION

BACKGROUND

1.1. The transport of radioactive material involves potential radiological hazards. To ensure the protection and safety of people, property and the environment, appropriate regulations, both at the national level and at the international level, are necessary. Government authorities regulate the transport of radioactive material by means of national regulations, in which the relevant international regulations and recommendations are taken into account. This Safety Guide provides recommendations for ensuring that the transport of radioactive material, both domestic and international, is conducted in compliance with IAEA Safety Standards Series No. SSR-6 (Rev. 1), Regulations for the Safe Transport of Radioactive Material, 2018 Edition [1] (hereinafter referred to as the 'Transport Regulations').

1.2. Compliance assurance is defined in the Transport Regulations as "a systematic programme of measures applied by a *competent authority* that is aimed at ensuring that the provisions of these Regulations are met in practice." Paragraph 307.2 of IAEA Safety Standards Series No. SSG-26 (Rev. 1), Advisory Material for the IAEA Regulations for the Safe Transport of Radioactive Material (2018 Edition) [2], states:

> "As used in the Transport Regulations, the term 'compliance assurance' has a broad meaning that includes all the measures applied by a competent authority that are intended to ensure that the requirements of the Transport Regulations are complied with in practice."

Therefore, essentially all activities of the competent authority are deemed part of its programme to ensure compliance with the Transport Regulations and are addressed in this Safety Guide. The recommendations provided in this Safety Guide supplement the recommendations provided in IAEA Safety Standards Series Nos GSG-12, Organization, Management and Staffing of the Regulatory Body for Safety [3], and GSG-13, Functions and Processes of the Regulatory Body for Safety [4].

1.3. IAEA Safety Standards Series No. GSR Part 2, Leadership and Management for Safety [5], establishes requirements for establishing, sustaining and continually improving leadership and management for safety, and an effective management system. This Safety Guide also uses the concept of a 'management system', which reflects and includes the initial concept of 'quality control' (controlling the quality of products) and its evolution through 'quality assurance' (the system for ensuring the quality of products) and 'quality management' (the system for managing quality).

1.4. Management systems implemented by users[1] of the Transport Regulations, as required by para. 306 of the Transport Regulations, are an important component of compliance assurance, and inspection[2] of these management systems by the competent authority is an effective means of monitoring the compliance of the user with the Transport Regulations.

1.5. This Safety Guide supersedes IAEA Safety Standards Series No. TS-G-1.5, Compliance Assurance for the Safe Transport of Radioactive Material[3].

OBJECTIVE

1.6. The objective of this Safety Guide is to assist competent authorities in the development and maintenance of compliance assurance programmes for the transport of radioactive material. This Safety Guide is intended to assist in ensuring a uniform application of the Transport Regulations by providing recommendations on the actions that competent authorities should perform in relation to their compliance assurance programmes.

1.7. This Safety Guide is intended to be used by competent authorities that are establishing or further developing programmes to ensure compliance with the Transport Regulations. The recommendations provided will also be useful to competent authorities with established compliance assurance programmes. Additionally, the Safety Guide will assist users of the Transport Regulations in their interactions with competent authorities.

---

[1] In this Safety Guide, a 'user' is a person who — or an organization associated with and involved in the movement of radioactive material that — designs, manufactures, maintains and repairs packagings, and prepares, consigns, loads, carries (including in-transit storage), ships after storage, unloads, receives, or otherwise uses a package in connection with the transport of radioactive material.

[2] In some Member States, the term 'audit' is used when referring to inspection activities related to a management system.

[3] INTERNATIONAL ATOMIC ENERGY AGENCY, Compliance Assurance for the Safe Transport of Radioactive Material, IAEA Safety Standards Series No. TS-G-1.5, IAEA, Vienna (2009).

SCOPE

1.8. This Safety Guide addresses compliance assurance for the safe transport of radioactive material, based on the same scope as described in paras 106–110 of the Transport Regulations.

STRUCTURE

1.9. Section 2 provides recommendations on the responsibilities and functions of the competent authority. Section 3 provides information on the various national and international regulations and guides for the transport of radioactive material. Section 4 provides recommendations on various aspects of the compliance assurance programme. Section 5 provides information on multilateral approvals. Section 6 provides recommendations on cooperation between competent authorities at the international level. The annexes provide examples of procedures and checklists for use by a competent authority in its compliance assurance programme.

2. RESPONSIBILITIES AND FUNCTIONS

REGULATORY BASIS

2.1. The government is required to assign the prime responsibility for safety to the person or organization responsible for a facility or an activity (which includes the transport of radioactive material) (see Requirement 5 of IAEA Safety Standards Series No. GSR Part 1 (Rev. 1), Governmental, Legal and Regulatory Framework for Safety [6]). The prime responsibility for ensuring safety in transport rests with consignors and carriers, who are required to take account of all relevant regulations.

2.2. Paragraph 307 of the Transport Regulations states that "The *competent authority* shall assure compliance with these Regulations." This compliance includes the oversight and enforcement of all regulations. In addition, certain activities of the competent authority are directly related to specific requirements of the Transport Regulations, such as the issuing of approvals and the allocation of identification marks.

2.3.   A State whose framework and arrangements for the transport of radioactive material are not yet fully established may develop its compliance assurance programme in stages, depending on the size of the transport industry. IAEA Safety Standards Series No. SSG-44, Establishing the Infrastructure for Radiation Safety [7], provides recommendations on the general approach to be followed. In an effective programme for compliance assurance, all users of the Transport Regulations and regulatory bodies (which may share responsibilities with the competent authority) should be considered.

ESTABLISHMENT OF A FRAMEWORK FOR SAFETY

2.4.   GSR Part 1 (Rev. 1) [6] establishes requirements for the roles and responsibilities of competent authorities[4]. Requirement 2 of GSR Part 1 (Rev. 1) [6] states that **"The government shall establish and maintain an appropriate governmental, legal and regulatory framework for safety within which responsibilities are clearly allocated."**

2.5.   Paragraph 2.5 of GSR Part 1 (Rev. 1) [6] states (footnote omitted):

"The government shall promulgate laws and statutes to make provision for an effective governmental, legal and regulatory framework for safety. This framework for safety shall set out the following:

(1)   The safety principles for protecting people — individually and collectively — society and the environment from radiation risks, both at present and in the future;
(2)   The types of facilities and activities that are included within the scope of the framework for safety;
(3)   The type of authorization that is required for the operation of facilities and for the conduct of activities, in accordance with a graded approach;
(4)   The rationale for the authorization of new facilities and activities, as well as the applicable decision making process;
(5)   Provision for the involvement of interested parties and for their input to decision making;

---

[4] The term 'competent authority' is used in the Transport Regulations for consistency with terminology used in the wider field of the regulation of the transport of dangerous goods. In the IAEA Safety Standards Series, the more general term 'regulatory body' is used, with which 'competent authority' is essentially synonymous.

(6) Provision for assigning legal responsibility for safety to the persons or organizations responsible for the facilities and activities, and for ensuring the continuity of responsibility where activities are carried out by several persons or organizations successively;

(7) The establishment of a regulatory body, as addressed in Requirements 3 and 4 [of GSR Part 1 (Rev. 1) [6]];

(8) Provision for the review and assessment of facilities and activities, in accordance with a graded approach;

(9) The authority and responsibility of the regulatory body for promulgating (or preparing for the enactment of) regulations and preparing guidance for their implementation;

(10) (Provision for the inspection of facilities and activities, and for the enforcement of regulations, in accordance with a graded approach;

(11) Provision for appeals against decisions of the regulatory body;

(12) Provision for preparedness for, and response to, a nuclear or radiological emergency;

(13) Provision for an interface with nuclear security;

(14) Provision for an interface with the system of accounting for, and control of, nuclear material;

(15) Provision for acquiring and maintaining the necessary competence nationally for ensuring safety;

……

(18) The specification of offences and the corresponding penalties;

(19) Provision for controls on the import and export of nuclear material and radioactive material, as well as for their tracking within, and to the extent possible outside, national boundaries, such as tracking of the authorized export of radioactive sources."

2.6. Paragraph 2.6 of GSR Part 1 (Rev. 1) [6] states that "Where several authorities are involved, the government shall specify clearly the responsibilities and functions of each authority within the governmental, legal and regulatory framework for safety."

2.7. The government is required to ensure that the competent authority is effectively independent in its safety related decision making and that it has functional separation from entities having responsibilities or interests that could unduly influence its decision making (see Requirement 4 of GSR Part 1 (Rev. 1) [6]).

LIST OF NATIONAL COMPETENT AUTHORITIES

2.8.   A list of national competent authorities for the safe transport of radioactive material (competent authorities responsible for approvals and authorizations related to the transport of radioactive material) is available on the IAEA–Global Nuclear Safety and Security Network web site.[5] Competent authorities should ensure that the information provided in this list is checked at least annually to verify that it is correct.

INTERLINKED RESPONSIBILITIES

2.9.   As stated in para. 207.1 of SSG-26 (Rev. 1) [2]:

"The competent authority is the organization defined by legislative or executive authority to act on behalf of a country in matters involving the transport of radioactive material, or an international authority on such matters. The legal framework of a country determines how a national competent authority is designated and is given the responsibility to ensure application of the Transport Regulations. In some instances, authority over different aspects of the Transport Regulations is assigned to different agencies, depending on the transport mode (air, road, rail, sea or inland waterway) and on the package and radioactive material type (excepted, industrial, Type A, Type B(U), Type B(M) and Type C packages; special form radioactive material, low dispersible radioactive material (LDRM), fissile material or uranium hexafluoride). A national competent authority may, in some cases, delegate the approval of package designs and certain types of shipment to another organization having the necessary technical competence. National competent authorities also constitute the competent authorities referred to in any conventions or agreements on the transport of radioactive material to which the country adheres."

2.10. Recommendations regarding the liaison between the competent authority and other governmental organizations are provided in paras 4.44–4.48 of GSG-12 [3].

---

[5]   https://gnssn.iaea.org/main/GlobalTransportNetworks/Pages/CompetentAuthorities.aspx

ORGANIZATION AND MANAGEMENT OF THE COMPETENT
AUTHORITY

2.11. GSG-12 [3] provides recommendations on meeting the requirements of
GSR Part 1 (Rev. 1) [6] in respect of the organizational structure, management
and staffing of the competent authority to support competent authorities in
performing their responsibilities and functions efficiently and effectively and in
an independent manner.

2.12. GSG-12 [3] provides recommendations on the following:

(a) The general characteristics of a competent authority with responsibility for
     safety (see paras 2.1–2.29 of GSG-12 [3]).
(b) Management for safety, focused on leadership for safety and safety culture
     aspects (see paras 3.1–3.26 of GSG-12 [3]).
(c) The organizational aspects necessary for the implementation of the core
     regulatory functions of a competent authority, including the following:
     (i)   Development and provision of regulations and guides (see paras 4.5
            and 4.6 of GSG-12 [3]);
     (ii)   Notification and authorization, including approvals (see paras 4.7–4.10
            of GSG-12 [3]);
     (iii) Regulatory review and assessment (see paras 4.11 and 4.12 of
            GSG-12 [3]);
     (iv)   Regulatory inspection (see paras 4.13–4.16 of GSG-12 [3]);
     (v)   Enforcement (see paras 4.17–4.20 of GSG-12 [3]);
     (vi)   Emergency preparedness and response (see paras 4.21 and 4.22 of
            GSG-12 [3]);
     (vii) Communication and consultation with interested parties (see
            paras 4.23 and 4.24 of GSG-12 [3]).
(d) The organizational aspects necessary to support the regulatory functions of
     a competent authority, including the following:
     (i)   Administrative support, including human resources, finance,
            management of documents and records, and equipment purchasing
            and control (see para. 4.26 of GSG-12 [3]);
     (ii)   Legal assistance (see paras 4.27–4.30 of GSG-12 [3]);
     (iii) Arrangements for contracting external expert support, where needed
            (see paras 4.40–4.43 of GSG-12 [3]);
     (iv)   International cooperation (see paras 4.49–4.54 of GSG-12 [3]).
(e) The characteristics of an integrated management system necessary for
     an effective and efficient competent authority (see paras 5.1–5.70 of
     GSG-12 [3]).

(f)   The necessary staffing and competences that should be in place to enable the competent authority to effectively perform its functions and to discharge its responsibilities (see paras 6.1–6.86 of GSG-12 [3]).

2.13. GSG-12 [3] should be read in conjunction with GSG-13 [4], which covers the technical aspects of the core functions of the competent authority and the processes by which they are discharged. The core functions of the competent authority interact with one another. For example, regulations and guides set out the regulatory requirements to be used in review and assessment, in the authorization or approval process, in inspections and in determination of enforcement actions. Similarly, the findings of reviews and assessments guide the approach to inspection, and inspection provides areas for review and assessment. Review and assessment, and inspection, may all influence the development of regulations and guides.

EXPERTISE AVAILABLE TO THE COMPETENT AUTHORITY

2.14. Requirement 3 of GSR Part 1 (Rev. 1) [6] states that "**The government… shall establish and maintain a regulatory body…and provide it with the competence and the resources necessary to fulfil its statutory obligation for the regulatory control of facilities and activities.**" Paragraph 2.36(b) of GSR Part 1 (Rev. 1) [6] states:

> "The government…[s]hall make provision for adequate arrangements for the regulatory body and its support organizations to build and maintain expertise in the disciplines necessary for discharge of the regulatory body's responsibilities in relation to safety".

The government should make arrangements for the competent authority to have access to expertise in many different fields, which could include the following:

(a)   Containment of the radioactive contents;
(b)   Criticality safety;
(c)   Radiation safety, including shielding analysis;
(d)   Thermal analysis;
(e)   Materials science and mechanical or structural engineering;
(f)   Management system;
(g)   Packaging testing;
(h)   Packaging manufacturing;

(i)   Packaging maintenance[6];
(j)   Packaging and transport operations;
(k)   Inspection and enforcement;
(l)   Emergency preparedness and response.

2.15. If all the necessary expertise is not available within the competent authority, external expert support should be sought. GSG-12 [3] provides recommendations on the following:

(a)   Developing and managing the competences of the competent authority staff, including through training;
(b)   Managing external expert support.

2.16. Owing to the international aspects of transport, attendance at international conferences and seminars is also important for the education and training of employees of the competent authority.

LIAISON BY THE COMPETENT AUTHORITY WITH OTHER GOVERNMENT AGENCIES

2.17. As noted in para. 2.9, more than one competent authority may be responsible for the regulatory control of transport in a State. For example, the following organizations may have roles and responsibilities concerning the safe transport of radioactive material:

(a)   Agencies with responsibilities for transport;
(b)   Agencies with responsibilities for dangerous goods;
(c)   Agencies with responsibilities for health and safety;
(d)   Agencies with responsibilities for radiation protection;
(e)   Agencies with responsibilities for emergency preparedness and response;
(f)   Law enforcement agencies;
(g)   Customs agencies;
(h)   Postal authorities;
(i)   National research institutes and institutes for materials testing;

---

[6] 'Maintenance' means the organized activity, both administrative and technical, of keeping structures, systems and components in good operating condition, including both preventive and corrective (or repair) aspects. Periodic maintenance is a form of preventive maintenance consisting of servicing, parts replacement, surveillance or testing at predetermined intervals based on calendar time, operating time or number of cycles [8].

(j)   Institutions that provide training and education.

2.18. The competent authority should facilitate regular cooperation between the parties within this complex network of agencies and persons to ensure the following:

(a)   An exchange of information regarding existing regulations for the transport of radioactive material;
(b)   An exchange of information on changes to national laws and regulations as well as changes to the Transport Regulations;
(c)   A complete programme of training for personnel at all levels;
(d)   Consistent application of inspection and enforcement relating to compliance assurance;
(e)   A regular review of all measures for emergency response, including the responsibilities of the competent authority, the industry for the transport of radioactive material and other relevant agencies;
(f)   A suitable forum for the discussion and resolution of issues relating to the Transport Regulations and compliance assurance;
(g)   An exchange of information regarding threats and vulnerabilities.

2.19. The appropriate competent authority should establish formal agreements with agencies whose purview may have an interface with the transport of radioactive material. Examples of such agencies are as follows:

(a)   Agencies that serve as the regulatory body for nuclear safety, radiation safety and radioactive waste safety, which includes the use and storage of radioactive material;
(b)   Agencies with responsibilities for transport;
(c)   Agencies with responsibilities for specific modes of transport;
(d)   National agencies involved in nuclear material accounting and control;
(e)   National agencies with responsibilities for the security of radioactive material (including nuclear material);
(f)   Customs agencies;
(g)   Environmental agencies;
(h)   Agencies with responsibilities for dangerous goods;
(i)   Agencies with responsibilities for radioactive or hazardous waste;
(j)   Agencies with responsibilities for emergency preparedness and response;
(k)   Other technical regulatory bodies;
(l)   Agencies with responsibility for law enforcement and the prosecution of criminal activities.

2.20. To ensure the security of radioactive material, nuclear material accounting and control, and physical protection of nuclear material, the requirements in some States necessitate that the competent authority maintain total control over all transit, import, export and inland shipments of radioactive material (including nuclear material). In such cases, applications in connection with the transfer of such material need to be checked by the competent authority prior to shipment to confirm that all proposed shipments and packages are in compliance with the Transport Regulations. Such checks are often required by national regulations, irrespective of whether the proposed shipment or package requires approval in accordance with the Transport Regulations. In such cases, the liaison between the supervising authorities should be extremely close.

2.21. In addition to the cooperation described in para. 2.18, the competent authority should be prepared to provide timely consultation for customs officers on the complex collection of documents that accompany shipments of radioactive material at national customs points. The competent authority should establish provisions to ensure the confidentiality of sensitive information related to the transport of nuclear material and radioactive material.

2.22. The competent authority should liaise very closely with agencies involved in emergency preparedness and response. In practice, the plans of such agencies usually concern response to accidents involving all dangerous goods or response to a nuclear or radiological emergency. Further recommendations on planning and preparing for response to an emergency involving radioactive material in transport, including situations in which a nuclear security event is confirmed to be the initiating event, are provided in IAEA Safety Standards Series No. SSG-65, Preparedness and Response for a Nuclear or Radiological Emergency Involving the Transport of Radioactive Material [9].

INTERFACES OF SAFETY WITH NUCLEAR SECURITY

2.23. Requirement 12 of GSR Part 1 (Rev. 1) [6] states:

> **"The government shall ensure that, within the governmental and legal framework, adequate infrastructural arrangements are established for interfaces of safety with arrangements for nuclear security and with the State system of accounting for, and control of, nuclear material."**

2.24. Paragraph 2.40 of GSR Part 1 (Rev. 1) [6] states that "Safety measures and nuclear security measures shall be designed and implemented in an integrated manner so that nuclear security measures do not compromise safety and safety measures do not compromise nuclear security."

2.25. Provisions concerning, inter alia, the physical protection of nuclear material in transport against sabotage and theft are contained in the Convention on the Physical Protection of Nuclear Material [10] and its amendment [11], which are legally binding on the parties thereto. The convention and its amendment apply to the international and domestic transport of nuclear material used for peaceful purposes.

2.26. Security provisions for the transport of radioactive material are provided in the United Nations Model Regulations [12] and in modal regulations [13, 14].

2.27. Recommendations for the physical protection of nuclear material are provided in IAEA Nuclear Security Series No. 13, Nuclear Security Recommendations on Physical Protection of Nuclear Material and Nuclear Facilities (INFCIRC/225/Revision 5) [15]. Recommendations for the security of radioactive material are provided in IAEA Nuclear Security Series No. 14, Nuclear Security Recommendations on Radioactive Material and Associated Facilities [16]. Furthermore, guidance on security in the transport of nuclear material and radioactive material is provided in IAEA Nuclear Security Series Nos 26-G, Security of Nuclear Material in Transport [17], and 9-G (Rev. 1), Security of Radioactive Material in Transport [18].

2.28. To meet Requirement 12 of GSR Part 1 (Rev. 1) [6], concerning interfaces of safety with arrangements for nuclear security, the competent authority should undertake activities including the following:

(a)   Liaise closely with the national agencies responsible for transport security;
(b)   Review proposed security measures for the transport of radioactive material to ensure that they do not compromise transport safety;
(c)   Incorporate, as appropriate, requirements on nuclear security mentioned in paras 2.25 and 2.26 into national requirements for the transport of radioactive material (including nuclear material), taking into account recommendations and guidance as described in para. 2.27;
(d)   If security arrangements are reviewed during inspections or as part of a review of an application for authorization by the competent authority, use available checklists and procedures that address security arrangements;
(e)   Ensure that its staff members receive training in nuclear security and are trustworthy, as appropriate, for their duties and responsibilities.

3. REGULATIONS AND GUIDES

3.1. Paragraph 3.3 of GSG-13 [4] states:

"The provision of regulations and guides is subject to Requirements 32–34 of [GSR Part 1 (Rev. 1) [6]]. The system of regulations and guides should be in accordance with the legal system of the State, and the nature and extent of the facilities and activities to be regulated. The regulations and guides should specify the requirements and associated criteria for ensuring the protection of people and the environment."

INTERNATIONAL AGREEMENTS AND GUIDES

3.2. The regulations and guides of a State should take into account that the transport of radioactive material is often international. National regulations as well as international modal regulations, which are based on the Transport Regulations, apply to such transport.

3.3. International bodies have issued general and modal recommendations and regulations on the safe transport of dangerous goods, including radioactive material (Class 7), as follows:

(a) Recommendations on the Transport of Dangerous Goods, Model Regulations ('United Nations Model Regulations' or 'The Orange Book') [12];

(b) Technical Instructions for the Safe Transport of Dangerous Goods by Air (ICAO-TI) [13], which amplify the basic provisions of annex 18 to the Convention on International Civil Aviation (Chicago Convention) [19];

(c) International Maritime Dangerous Goods Code (IMDG Code) [14], which provides detailed regulations for the carriage of dangerous goods in packaged form by sea under chapter VII of the International Convention for the Safety of Life at Sea (SOLAS) [20];

(d) Regulations to the Universal Postal Convention, which provide detailed requirements for the exceptional sending of dangerous goods (e.g. radioactive material) by international postal service based on the provisions of the Universal Postal Convention [21].

The United Nations Model Regulations, ICAO-TI and the IMDG Code are updated every two years. Provisions for Class 7 goods (radioactive material) contained in the United Nations Model Regulations are based on the IAEA

Transport Regulations and are incorporated into the ICAO-TI and the IMDG Code thereafter. The application of the ICAO-TI for air transport, the IMDG Code for sea transport and the Regulations to the Universal Postal Convention for international postal service are mandatory for States that are party to those conventions.

3.4. There are also regional agreements, conventions and regulations concerning the safe transport of dangerous goods, including radioactive material, which may be mandatory for States that are party to these agreements, conventions and regulations. Examples of such agreements, conventions and regulations that take due account of the Transport Regulations include the following:

(a) The Agreement Concerning the International Carriage of Dangerous Goods by Road (ADR) [22];
(b) The Convention Concerning International Carriage by Rail [23] and its appendix C, the Regulations Concerning the International Carriage of Dangerous Goods by Rail [24];
(c) The European Agreement Concerning the International Carriage of Dangerous Goods by Inland Waterways [25];
(d) The MERCOSUR/MERCOSUL Agreement of Partial Reach to Facilitate the Transport of Dangerous Goods, signed by the Governments of Argentina, Brazil, Paraguay and Uruguay in 1994 [26];
(e) The Agreement on International Goods Traffic by Rail [27] and its annex 2.

3.5. In the interests of international harmonization and safety, individual States should participate in the relevant international and regional conventions and agreements mentioned in paras 3.3 and 3.4 and follow and fully implement the provisions of the Transport Regulations. However, owing to specific national circumstances, a State may need to deviate from, or add to, the provisions of the Transport Regulations or of other international regulations and guidelines. In such cases, these specific provisions should be included in relevant national regulations and guides and, if applicable, in international or regional regulations (e.g. under 'State Variations' in ICAO-TI [13]); if practicable, the competent authority should communicate such differences to relevant transport organizations, to other competent authorities as appropriate, and to the international modal organizations.

NATIONAL REGULATIONS AND GUIDES

3.6. In accordance with Requirements 32–34 of GSR Part 1 (Rev. 1) [6], the competent authority is required to establish or adopt regulations and guides.

National regulations and guides for the transport of radioactive material should be appropriate to the size and type of transport industry to which they apply.

3.7. GSG-13 [4] acknowledges that the development and provision of regulations and guides is a core function of a competent authority and provides recommendations on this function. After describing the objectives of regulations and guides and their differences, GSG-13 [4] provides recommendations on the following:

(a) The key principles to be followed for the initial development and subsequent review and revision of regulations and guides, as necessary, including the involvement of interested parties and the consideration of international or national standards and operating experience, as well as developments in transport practices;
(b) Topics that should be addressed in regulations and guides, for example the authorization process, the documentation to be submitted to the competent authority, and the enforcement policy;
(c) The promotion of regulations and guides to interested parties.

3.8. National regulations for the safe transport of radioactive material should be based on the Transport Regulations; furthermore, in the preparation of national regulations and guides for the transport of radioactive material, all relevant international agreements, regulations and recommendations should be taken into account. The language used in the preparation of such regulations and guides should be appropriate to ensure correct and unambiguous understanding by the users of the regulations. If international regulations and/or modal conventions are adopted or used as national regulations, they should be translated into the official national language(s) and the accuracy of the translations should be verified. Recommendations on maintenance of national regulations are provided in para. 4.95.

4. COMPLIANCE ASSURANCE

4.1. The competent authority should establish a programme for compliance assurance that applies to all relevant aspects of the transport of radioactive material within its jurisdiction or area of influence with regard to safety and the provisions of the Transport Regulations.

4.2. Compliance with the Transport Regulations can be ensured by the competent authority in various ways and may include the following activities:

(a)   Issuing of approvals;
(b)   Inspections of the management systems of users of the Transport Regulations;
(c)   Training and distribution of information;
(d)   Assessment of designs;
(e)   Inspection of testing, which should include the direct observation of specific tests;
(f)   Inspection of transport operations, which should include direct observation of transport activities;
(g)   Inspection of manufacturing, which should include direct observation of specific steps in the manufacturing process;
(h)   Inspection of maintenance arrangements, which should include direct observation of maintenance activities;
(i)   Emergency preparedness and response;
(j)   Enforcement actions and investigations of incidents;
(k)   Regular review of the national legal framework, including national and international regulations;
(l)   Liaison and cooperation with other government agencies (see paras 2.17–2.22).

A graphic representation of these activities is provided in Fig. 1 in the form of the 'compliance assurance circle'.

4.3. These activities are not meant to be implemented in any particular order, and a national compliance assurance programme might not necessarily cover all these activities. The extent of the programme depends on the quantities and types of package being transported and on the size and complexity of the transport industry for which the competent authority has responsibility, as well as its own resources.

4.4. In all circumstances, a competent authority's compliance assurance programme should include, at a minimum, the following activities:

(a)   Activities relating to review and assessment, including the issuing of certificates of approval;
(b)   Activities relating to inspection and enforcement;
(c)   Activities relating to emergency preparedness and response.

FIG. 1. Compliance assurance circle.

DEVELOPMENT AND IMPLEMENTATION OF A COMPLIANCE ASSURANCE PROGRAMME

4.5. The steps in the initial development of a compliance assurance programme can be summarized as follows:

(1) Determination or confirmation of the size and state of the existing industry for the transport of radioactive material;
(2) Determination or confirmation of the existing legal powers, independence and other resources available to the competent authority;
(3) Establishment of liaison with government agencies or organizations having a legitimate interest in or an interface with aspects of the transport of radioactive material;
(4) Provision of a sound legal framework to enable the effective functioning of the competent authority;

(5) Formal confirmation of the working relationships between other government agencies and other organizations in respect of the transport of radioactive material;

(6) Gathering of further detailed information on the size of the industry for the transport of radioactive material, including information on package types and number of movements;

(7) Formal specification of the size, structure and resources of the competent authority and development of a management system for the competent authority;

(8) Initial training of personnel of the competent authority and other personnel involved in the enforcement of regulations;

(9) Creation or adoption of national regulations for the transport of radioactive material (with provision for all package types, transport operations and modes of transport);

(10) Development and implementation of an initial compliance assurance programme;

(11) Distribution of information to all parts of the industry regarding the competent authority's policies, regulations and guides for the transport of radioactive material;

(12) Collection of initial evidence of compliance with the Transport Regulations by means of the applicable activities described in para. 4.2;

(13) Accumulation and review of evidence of compliance on a continual basis.

4.6. Compliance assurance programmes may be relatively simple and straightforward or may be complex and wide ranging, commensurate with the size and variety of the transport industry for which the competent authority has responsibility. At a minimum, for a simple compliance assurance programme in a State that performs a limited number of shipments including only a few types of radioactive material, account should be taken of the following:

(a) Radioactive material classification;
(b) Import and export operations;
(c) All relevant modes of transport;
(d) All relevant package types;
(e) Associated package certificates of foreign origin, if applicable;
(f) Maintenance and removal from service of a packaging.

4.7. A more complex compliance assurance programme will be needed for a State where a large number of shipments occur within, through, into or from its territory, including shipments of many types and large quantities of radioactive material, or for a State where packages are designed and manufactured. For such a programme, account should additionally be taken of package design, testing, manufacture and approval.

4.8. The activities stated in para. 4.4 should be addressed in a manner that is graded in accordance with the complexity and variety of the particular responsibilities of the competent authority. Recommendations on applying a graded approach to the functions and processes of a competent authority are provided in section 2 of GSG-13 [4].

4.9. After a compliance assurance programme has been developed and introduced, it should be reviewed periodically by the competent authority to take account of regulatory changes and the experience of users of the Transport Regulations. The compliance assurance programme should be updated in a timely manner when there are changes to the Transport Regulations and should be reviewed periodically to ensure that it continues to achieve the goals it was designed to achieve. In some cases, such reviews may be performed by qualified external organizations.

ISSUING OF APPROVALS BY THE COMPETENT AUTHORITY

4.10. The Transport Regulations distinguish between cases in which radioactive material can be transported without approval by the competent authority and cases in which approval is required. In all cases, the Transport Regulations place the primary responsibility for compliance on the consignor. The following require competent authority approval (see para. 802 of the Transport Regulations), and for each of these an appropriate independent assessment should be made by the competent authority:

(a) Designs for:
    (i) Special form radioactive material;
    (ii) Low dispersible radioactive material;
    (iii) Fissile material excepted under para. 417(f) of the Transport Regulations;
    (iv) Packages containing 0.1 kg or more of uranium hexafluoride;
    (v) Packages containing fissile material, unless excepted by paras 417(a–f), 674 or 675 of the Transport Regulations;
    (vi) Type B(U) packages and Type B(M) packages;
    (vii) Type C packages.
(b) Special arrangements.
(c) Certain shipments.

(d)   Radiation protection programmes for special use vessels.

(e)   Calculation of radionuclide values not listed in table 2 of the Transport Regulations.

(f)   Calculation of alternative activity limits for an exempt consignment of instruments or articles.

As described in section VIII of the Transport Regulations, some of the items listed above may be subject to multilateral approval (i.e. the approval of several competent authorities).

4.11. In accordance with paras 803, 807(b) and 808 of the Transport Regulations, unilateral approval by the competent authority of the country of origin of the design is required for the following:

(a)   The design of special form radioactive material;

(b)   The design of packages containing 0.1 kg or more of uranium hexafluoride that meet the requirements of paras 631–633 of the Transport Regulations;

(c)   Type B(U) and Type C package designs, except package designs for fissile material and Type B(U) package designs for low dispersible radioactive material.

4.12. In accordance with paras 403(a), 803, 805, 807(a), 808(a), 808(b), 811, 814, 817, 820(a), 820(b), 825 and 829 of the Transport Regulations, multilateral approval by the competent authority is required for the following:

(a)   The determination of basic radionuclide values not listed in table 2 of the Transport Regulations;

(b)   The design of low dispersible radioactive material;

(c)   The design of material excepted from fissile classification under para. 417(f) of the Transport Regulations;

(d)   The design of packages containing 0.1 kg or more of uranium hexafluoride that meet the requirements of para. 634 of the Transport Regulations;

(e)   The design of packages for fissile material that are not excepted by any of the paras 417(a)–(f), 674 and 675 of the Transport Regulations;

(f)   The design of Type B(U) packages for low dispersible radioactive material;

(g)   The design of Type B(M) packages;

(h)   Alternative activity limits for an exempt consignment of instruments or articles;

(i)   Package designs approved by the competent authority under the provisions of the 1985 Edition or the 1985 Edition (As Amended 1990) of the Transport Regulations;

(j)   After 31 December 2025, package designs approved by the competent authority under the provisions of the 1996 Edition, 1996 Edition (Revised), 1996 (As Amended 2003), 2005, 2009 and 2012 Editions of the Transport Regulations;

(k)   The shipment of Type B(M) packages not conforming with the requirements of para. 639 of the Transport Regulations or designed to allow controlled intermittent venting;

(l)   The shipment of Type B(M) packages containing radioactive material with an activity greater than $3000A_1$ or $3000A_2$, as appropriate, or 1000 TBq, whichever is lower;

(m)   The shipment of packages containing fissile material, if the sum of the criticality safety indexes of the packages in a single freight container or in a single conveyance exceeds 50, except for shipments excluded from this requirement in accordance with para. 825(c) of the Transport Regulations;

(n)   Radiation protection programmes for shipments by special use vessels in accordance with para. 576(a) of the Transport Regulations;

(o)   The shipment of SCO-III (surface contaminated object of the group SCO-III);

(p)   Consignments transported under special arrangement.

Further recommendations on multilateral approvals are provided in Section 5.

4.13. It is the responsibility of the applicant to demonstrate compliance with the Transport Regulations, and it is the responsibility of the competent authority to review and assess compliance. Applicants should be encouraged to contact the competent authority during the preliminary design stages to discuss the implementation of the relevant design principles and the approval process. Informal and formal discussions may be held between applicants and the competent authority on acceptable ways of demonstrating compliance. Additionally, in some cases it is advantageous for both the prospective applicant and the competent authority to discuss an outline of the proposed application before it is formally submitted in detailed form.

4.14. The decision to grant an approval is based on the competent authority's evaluation of the applicant's demonstration of compliance with the relevant requirements of the Transport Regulations. Depending on the type of approval, the corresponding application should contain at least the information described in section VIII of the Transport Regulations (paras 803, 805, 807(c), 809, 812, 815, 817, 827, 827A and 830). Guidance on information to be included in applications for approval is provided in Annex I.

4.15. Upon receipt of an application for approval, the competent authority should evaluate whether all relevant regulatory requirements have been met. A list of items that should be included in an application is provided in Annex I to this Safety Guide. Paragraphs 3.147–3.209 of GSG-13 [4] provide recommendations on the review and assessment process, including the associated documentation. If the competent authority determines that the application demonstrates compliance with the Transport Regulations, the competent authority is required to provide the applicant with a certificate of approval containing all the necessary information (see section VIII of the Transport Regulations and Annex II to this Safety Guide).

4.16. When considering applications for approval of shipments under special arrangement, the competent authority should assess the demonstration by the applicant that the overall level of safety provided by the design of the package and by compensatory measures such as operational controls during transport is at least equivalent to that which would be achieved if all applicable regulatory requirements were met. Possible additional operational controls that might be employed are addressed in para. 830.1 of SSG-26 (Rev. 1) [2].

4.17. Whenever possible, standard formats should be used for each type of certificate. Minimum requirements for the content of certificates of approval are specified in paras 834–839 of the Transport Regulations. Example templates for certificates of approval for use by the competent authority are provided in Annex II to this Safety Guide.

4.18. Consistent with the national practice and with regard for commercial considerations, the competent authority should supply copies or provide information on its approvals to other competent authorities and users of the Transport Regulations to facilitate compliance with any specific requirements or conditions. Depending on national practice, information on certificates of approval for designs used for transport may be provided on the applicable competent authority web site.

THE MANAGEMENT SYSTEM OF USERS OF THE TRANSPORT REGULATIONS IN SUPPORT OF COMPLIANCE ASSURANCE

4.19. To meet the requirements of para. 306 of the Transport Regulations and Requirements 3–8 of GSR Part 2 [5], an integrated management system is required for all transport related activities. The extent of the management system will depend on the types of transport activity being considered (e.g. design, testing, manufacture, use, maintenance), ranging from a relatively simple system for the infrequent transport of packages that do not require approval by the competent authority to a more complex system for the regular transport of packages subject to such approval. Annex I to IAEA Safety Standards Series No. TS-G-1.4, The Management System for the Safe Transport of Radioactive Material [28], provides information on how to address the various elements of the management system.

4.20. As stated in para. 306 of the Transport Regulations, "Where *competent authority approval* is required, such *approval* shall take into account, and be contingent upon, the adequacy of the *management system.*" The competent authority should confirm the adequacy of the applicant's management system by reviewing the management system documentation and by inspecting the implementation of arrangements in practice (see para. 4.23). An example of the types of information that the competent authority may consider in determining the adequacy of a management system as a pre-condition for issuing approvals is provided in Annex I. If the competent authority has confirmed the existence of a satisfactory management system, it may issue a 'certificate of compliance' for the management system, if required by national regulations.

4.21. The competent authority should establish an inspection programme to verify that the user's management system covers all the relevant aspects identified in TS-G-1.4 [28] and is implemented and followed correctly. This should include management systems implemented in the transport of packages that are not subject to competent authority approval. A list of items that the competent authority may consider during inspections of the management systems of users and related inspection activities is provided in Annex III. An example procedure and an example checklist for inspecting a management system are provided in Annexes IV and V, respectively.

4.22. Irrespective of the size of the organization concerned or the scale of its activities, the competent authority should verify through inspections that, consistent with the recommendations provided in TS-G-1.4 [28], the management system of the user comprises at least the following:

(a) An organizational structure and competent personnel suitable for administering and conducting activities in the management system;
(b) The capability to develop all procedures and instructions needed to guide, control and verify the conduct and evolution of activities in the management system;
(c) The means to develop, maintain and make accessible to the competent authority all necessary records and documentation of the management system;

(d)   The performance of activities to ensure compliance with the Transport
        Regulations and any additional national requirements.

4.23. In verifying the effectiveness of the arrangements within the management
system of a user, the competent authority should inspect procedures, records and
facilities, especially facilities in which designers and manufacturers perform their
operations. The competent authority should verify the following, as appropriate:

(a)   The design of a package is accurately described by engineering drawings,
        material specifications and records of the methods of construction. For
        package designs requiring approval by the competent authority, this
        information is a required part of the application for the certificate of approval
        (section VIII of the Transport Regulations). For package designs that do
        not require approval by the competent authority, the information should be
        provided to the competent authority upon request.
(b)   The packagings are manufactured in accordance with the design. For
        package designs that require approval by the competent authority, changes or
        modifications in the construction methods for the packaging or the materials
        of construction are required to be approved by the competent authority
        before use of the package (see para. 503 of the Transport Regulations). For
        package designs that do not require approval by the competent authority,
        such changes should be documented and made available to the competent
        authority upon request. This applies equally to new package designs and to
        packagings in use.
(c)   Equipment used for inspection, measurement, testing and manufacturing
        is suitable for its purposes and is properly controlled, calibrated, used and
        maintained in accordance with procedures and schedules. All results from
        inspections, measurements and testing and all products of manufacturing
        should be fully documented.
(d)   The packages are correctly prepared, packed and transported. This includes
        all necessary maintenance and other administrative procedures, as well as
        appropriate measures for radiation protection.
(e)   All non-conformances are correctly documented and reviewed, and accepted
        or rejected, and the competent authority is notified, as appropriate.

TRAINING AND DISTRIBUTION OF INFORMATION

4.24. The competent authority should ensure through its compliance assurance
programme and its monitoring of management systems that all training needs
of the organizations involved in transport are identified and implemented as

required by paras 311–315 of the Transport Regulations. The training programme for an individual may be adjusted on the basis of the relevant experience and responsibilities of the person. The competent authority should, as appropriate, specify and participate in the training of persons involved in the transport of radioactive material. The competent authority should also specify and participate in the training of its own staff.

4.25. In accordance with para. 314 of the Transport Regulations, each organization is required to maintain adequate records of the training provided. These records should include the performance of individual trainees and the authorizations or certificates issued. Also, records should be maintained in accordance with the recommendations for the management system provided in TS-G-1.4 [28], and these records should be inspected periodically by the competent authority. The main purposes of such records are as follows:

(a) To provide evidence of the appropriate qualification of persons whose duties have a bearing on safety, including evidence of the required authorizations or certificates;
(b) To provide evidence of the basis for these authorizations or certificates;
(c) To provide documentation that can be used in reviews of the training programme to enable any necessary corrective actions to be taken.

In some States, the holders of certain posts within the competent authority and the organizations of the consignor, carrier and/or consignee have to be authorized or certified before they are allowed to perform their duties.

4.26. Training material applicable to personnel involved in the transport of radioactive material is provided in Ref. [29] and on the IAEA's e-learning site[7].

4.27. The preparation and distribution of information and guidance by the competent authority is necessary for the implementation of a compliance assurance programme. Such information may be in the form of bulletins on important safety related matters. It may also be in the form of information notices and guides intended to assist users in the application of the Transport Regulations. It may also be provided through seminars, conferences or training courses for personnel of regulatory bodies, consignors, carriers and other groups to explain the correct application of the Transport Regulations. Further guidance concerning

---

[7] The IAEA provides a selection of e-learning materials on a wide range of topics, including topics discussed in this publication. These can be found on the IAEA's e-learning site at: https://elearning.iaea.org.

communication and consultation with interested parties by the competent authority can be found in IAEA Safety Standards Series No. GSG-6, Communication and Consultation with Interested Parties by the Regulatory Body [30].

ASSESSMENT OF DESIGNS

4.28. Paragraph 220 of the Transport Regulations states:

> "*Design* shall mean the description of *fissile material* excepted under para. 417(f), *special form radioactive material, low dispersible radioactive material, package* or *packaging* that enables such an item to be fully identified. The description may include specifications, engineering drawings, reports demonstrating compliance with regulatory requirements, and other relevant documentation."

Therefore, design is much more than the drawings and specifications that enable the packaging to be manufactured. The design to be assessed includes the supporting reports and documents that substantiate or verify statements or assumptions made by the designer. It also includes instructions for package preparation, instructions for maintenance and procedures for repair or modification.

4.29. Section VI of the Transport Regulations establishes requirements for special form radioactive material, low dispersible radioactive material, material excepted from fissile classification, and packagings and packages. In the case of the designs specified in para. 802 of the Transport Regulations, approval by the competent authority is required and hence the assessment of the design conducted by the competent authority should take account of the requirements in section VI of the Transport Regulations.

4.30. The design assessment conducted by the competent authority should consider any aspect of the design that could adversely affect one or more of the following:

(a)   Containment of the radioactive contents;
(b)   Control of external dose rate;
(c)   Prevention of criticality;
(d)   Prevention of damage caused by heat.

4.31. Where a number of very similar package designs exist, the assessor may make comparisons relating to the final acceptability of the designs; however, this should be done only after the detailed differences between the package designs have been identified by the applicant and confirmed by the competent authority as being of minor significance.

4.32. The assessment of the designs of packages intended to be used for shipment after storage should consider the effects of ageing mechanisms during an extended time period between loading of the package and its shipment after storage to ensure that the package design meets all applicable requirements of the relevant provisions of the Transport Regulations at the time when the first shipment after storage takes place (see paras 503(e) and 613A of the Transport Regulations). This includes the assessment of appropriate ageing management and a gap analysis programme (see para. 809(f) and (k) of the Transport Regulations). More guidance is provided in paras 503.3, 613A.1–613A.6, 809.3 and 809.4 of SSG-26 (Rev. 1) [2].

TESTING OF PACKAGES AND MATERIALS

4.33. It may be necessary to test packages and scale models or representative examples of package features and materials (including special form radioactive material) to demonstrate compliance of the design with the requirements in section VII of the Transport Regulations. Testing may be undertaken by the designer, the applicant, a third party testing organization or the competent authority. The following points should be considered when determining compliance with the requirements for testing:

(a) The organization performing the test should have an appropriate management system that addresses all aspects of the testing. It should cover not only the manufacture of the specimens to be tested but also all relevant activities relating to management, preparation, measuring, testing, recording, analysing (including corrective measures, if necessary) and reporting associated with the particular test or series of tests to be undertaken.

(b) The test programme should satisfy the approving body (i.e. the competent authority or other appropriate organization). The number of tests and specimens, the test conditions, the test sequences, the measurement techniques, the methods of analysis and the acceptance criteria should be clearly established. When drop tests are included, drop sequences and drop attitudes should be agreed with the approving body. Owing to unexpected results during testing, some variation in the test programme may be necessary in the course of testing, and allowance should be made for this fact when preparing test specimens, scheduling tests and using test facilities.

(c)   The objectives and parameters of the tests should be clearly established. It should be made clear whether (i) the sole objective of the tests is a straightforward verification that the package meets all the requirements of the Transport Regulations or only some of these requirements, (ii) the designer wants different (e.g. more stringent) test criteria to be applied or (iii) additional information is being sought from the tests to improve the designer's knowledge of the design principles, safety margins and performance.

(d)   It should be clearly established that the test facilities comply with the requirements of the Transport Regulations, particularly in the case of the targets used for drop and penetration tests, in which the weight of the test specimen is limited by the capacity of the test facility.

(e)   All measuring and monitoring equipment used before, during and after the tests to confirm and record the state of the test specimen and any forces imposed on it as a result of the tests, should be operated within the applicable national or international limits for the particular pieces of equipment. It should be verified that this equipment works accurately, within applicable national or international limits. This should be achieved by using properly calibrated measuring or test equipment, such as pressure and leak test equipment, accelerometers, strain gauges and thermal measuring apparatuses.

(f)   Adequate methods of recording the information obtained during the test programme should be implemented, and appropriate test records should be made available to the competent authority so that compliance with the requirements of the Transport Regulations can be confirmed.

(g)   All test results, including any instances of damage, should be considered as part of the competent authority's assessment of the final package design.

4.34. When a scale model is used in testing to support an application for approval, the competent authority should confirm that all scaling factors have been taken into account, with all pertinent features of the package design being accurately represented.

4.35. When conducting the final design assessment, the competent authority should take into account the packages tested, the test results, and any changes made to the package design after testing that were submitted by the applicant or designer in relation to the final design.

SPECIAL FORM RADIOACTIVE MATERIAL AND LOW DISPERSIBLE
RADIOACTIVE MATERIAL

4.36. The competent authority should determine whether the management system
arrangements for the design, testing and manufacture of special form radioactive
material or low dispersible radioactive material are appropriate and adequate for
the nature of the material and the amounts likely to be produced.

4.37. Before the commencement of tests by the applicant, the competent authority
should consider inspecting the test facilities and arrangements, especially the
specimens, the target for drop tests, and the measuring and recording systems. The
competent authority may also perform inspections that include direct observation
of the tests. The competent authority should require that it be informed by the
applicant about any deviation from the test plan and the resulting consequences.

4.38. The competent authority should verify that applications for approval of the
design of special form radioactive material or low dispersible radioactive material
include the test programme, the test results and the information described in
Annex I. The application should demonstrate that the regulatory requirements
have been met.

4.39. The competent authority should give consideration to the necessary
identification of special form radioactive material or low dispersible radioactive
material, as well as to the in-service inspections and safety checks to be made
to ensure the continued integrity of the special form radioactive material or low
dispersible radioactive material.

4.40. When the competent authority has verified through its own design
assessment (see paras 4.28–4.31, 4.41 and 4.52) that the design for special form
radioactive material or low dispersible radioactive material meets all the applicable
requirements, the competent authority is required to issue a certificate of approval
that attributes to the approved design an identification mark, in accordance with
para. 804 of the Transport Regulations. Example templates for certificates of
approval are provided in Annex II.

PACKAGES NOT REQUIRING APPROVAL BY THE COMPETENT
AUTHORITY

4.41. It is the responsibility of the competent authority to ensure that the
designs of packages are assessed against all the relevant parts of the Transport

Regulations. Therefore, the competent authority should conduct assessments of designs specified in para. 802(a) of the Transport Regulations and should ensure that similar assessments of package designs that do not require approval by the competent authority (e.g. Type A packages, industrial packages) are performed by appropriate organizations and that the necessary documentary evidence of such assessments is made available to the competent authority, if requested. (See paras 801.1–801.3 of SSG-26 (Rev. 1) [2] and IAEA Safety Standards Series No. SSG-66, Format and Content of the Package Design Safety Report for the Transport of Radioactive Material [31], for guidance on documentary evidence for packages that do not require competent authority approval.) Information about the structure and contents of a package design safety report, which applies to all types of package and is intended to demonstrate compliance of the design of a package with the Transport Regulations, is provided in SSG-66 [31].

4.42. The compliance assurance programme of the competent authority should also cover, through a graded approach, the design, manufacture and use of packages, and the maintenance of packagings, that do not require approval by the competent authority.

4.43. The competent authority should verify that the user complies with the requirements in paras 306 and 801 of the Transport Regulations for package designs that do not require approval by the competent authority. In particular, the following subjects for inspection by the competent authority should be addressed:

(a)　The management system under which the package is designed, manufactured and transported;
(b)　The design process and the internal process to provide documentary evidence that the package design meets all applicable requirements (see para. 4.41);
(c)　Control of manufacturing;
(d)　The programme for maintenance of packagings (in the case of reusable packagings).

PACKAGES REQUIRING COMPETENT AUTHORITY APPROVAL

4.44. The competent authority may discuss the development and the proposed testing of a package with the applicant on the basis of the preliminary information provided. The format and content of this preliminary information should take into account the recommendations provided in SSG-66 [31]. Specifically, the preliminary information might include the test plan for the package, with a clear statement of the scale of the model, the requirements and specifications of the model, the number of tests proposed, the drop attitudes for packages, the essential measuring and recording equipment to be used and the nature of the target for drop tests. The preliminary information might also cover the requirements of the management system for design, manufacture and testing.

4.45. The competent authority should consider special features of the package design, as well as the testing plan. If the applicant proposes to use a scale model specimen, it should be ensured that all relevant features of the original are adequately scaled and represented, including materials, contents and internal structures. The adequacy of the means proposed to establish compliance with the Transport Regulations should be reviewed. Account should be taken of the instrumentation to be used for the measurement of physical quantities such as local accelerations, strains and internal pressure transients.

4.46. The competent authority should verify that the manufacture of models or prototypes is undertaken in a controlled manner that complies with the management system used by the manufacturer so that the models or prototypes are representative of the proposed package design. Particular consideration should be given to the materials used, welding and inspections, the results of quality control tests, and any deviations from the requirements and specifications.

4.47. Before the commencement of tests by the applicant, the competent authority should consider inspecting the test facilities and arrangements, especially the specimen, the target for drop tests, and the measuring and recording systems. The competent authority may also perform inspections that include direct observation of the tests.

4.48. In conjunction with the information in Annex I and in SSG-66 [31] about the structure and contents of a package design safety report, the application for approval should include the test programme, the results of testing and the evaluation report (see also paras 4.33 and 4.35). The application should describe the management system of the applicant and should state the requirements for the production of packagings and their proper maintenance and use. The applicant should demonstrate that the requirements for the package type have been met. Specifically, the following aspects should be included, if appropriate, for routine, normal and accident conditions of transport:

(a) Containment of the radioactive contents;
(b) Control of external dose rate;
(c) Prevention of criticality;
(d) Prevention of damage caused by heat.

The application for approval should demonstrate compliance with the performance standards required by section VI of the Transport Regulations, including specific test requirements, using the methods listed in para. 701 of the Transport Regulations.

4.49. When assessing safety, the competent authority should, as appropriate, make independent assessments to verify the results presented in the application for approval of the package design, which includes the design assessment (see paras 4.28–4.32, 4.41 and 4.52). In making these independent assessments, the competent authority should ensure that proper computer codes, methods of calculation and models have been used; that they have been validated; and that all input data have been correctly and, if appropriate, conservatively defined.

4.50. When assessing applications for approval of package design, the competent authority should ensure that full and proper provision has been made for the legibility, durability and application of identification marks and serial numbers. This is particularly important in cases where multiple or interchangeable packaging components are used.

4.51. The design of a package should be accepted or rejected on the basis of the results of the evaluation. In the case of acceptance of the proposed design by the competent authority, a certificate of approval is required to be issued. Example templates for certificates of approval are provided in Annex II.

4.52. Reviews and assessments by competent authorities of applications for approval usually involve extensive resources, skills and expertise. The following aspects should be considered:

(a)   The assessor should have a thorough knowledge of the Transport Regulations pertinent to the design under assessment to ensure that the design will produce a package that is safe under routine, normal and accident conditions of transport.

(b)   Before commencing the design assessment, the assessor should be satisfied that a management system at an appropriate level has been applied throughout the design process; appropriate evidence of this should be made available to the assessor (paras 807(c), 809(j) and 815 of the Transport Regulations). Before initiating a detailed review of an application, the completeness of the design description and specification should be confirmed, including information regarding the intended use of the package design (e.g. mode of transport, handling during transport, transport after storage).

(c)   The assessor should thoroughly examine the thermal aspects of the package design; the assessor should consider both dissipation and absorption of heat in routine, normal and accident conditions of transport. Thermal stresses should be analysed to ensure that the leaktightness or mechanical properties of the package are not unduly compromised in routine and normal conditions of transport or in thermal test conditions.

(d)   The assessor should ensure that any computer codes, methods of calculation, or models used by the applicant are appropriate and have been validated, and that all input data have been correctly — and, if appropriate, conservatively — defined.

(e)   The assessor should examine all relevant mechanical aspects of the design to confirm that the package will be physically able to safely carry the specified radioactive material under routine, normal and accident conditions of transport (this includes, for example, tie-down points and trunnions). The assessor should analyse the structural attributes of the package and should verify that any impact or other damage the package might sustain in routine, normal or accident conditions of transport will not compromise its ability to meet the requirements of the Transport Regulations.

(f)   The assessor should examine all materials intended for use in manufacturing the package with regard to their correct specification and condition, their ability to perform satisfactorily under all expected and specified environmental conditions (e.g. temperature, pressure, irradiation, humidity) and their compatibility with other materials used.

(g)   The assessor should verify that ageing mechanisms have been taken into account.

(h)   The assessor should examine in detail the shielding features and radiation safety aspects of the design; the assessor should confirm that, with regard to the maximum proposed radioactive contents, the design of the package provides sufficient radiological shielding in all directions to comply with the Transport Regulations and the principle of optimization of protection. The assessor should confirm that any material used for shielding is physically and chemically stable and is not likely to move or to deteriorate during transport, since this would decrease the degree of shielding provided by the packaging. The absence of any radiation 'shine paths' through package closures and ports used for package testing should be verified.

(i)   The need to decontaminate the packagings in use should also be considered. The assessor should confirm the absence of features that might retain contamination and the absence of materials that are difficult to decontaminate.

(j)   The assessor should thoroughly examine all aspects of containment provided by the package. The assessor should also consider the features of the design that provide for containment and should determine how they might be adversely affected by routine and normal conditions of transport, by the prescribed maintenance periods and instructions, and by the effects of accident conditions of transport and related testing.

(k)   For packages designed to contain fissile material, the assessor should thoroughly examine all aspects of the package that are designed to maintain subcriticality during routine, normal and accident conditions. The assessor should specifically consider the contingencies listed in para. 673(a) and the requirements of para. 673(b) of the Transport Regulations.

(l)   The assessor should examine the in-service handling and use, inspection, and maintenance instructions and specifications in sufficient depth to verify that all such instructions and specifications are appropriate for the package as designed. The assessor should verify that in-service instructions and specifications provide for authorized repairs and modifications of the packaging. The procedures for repair and modification should be agreed with the assessor. The assessor should also consider that such package instructions may have to be followed by carriers and consignees that are unfamiliar with the package and its design principles.

IDENTIFICATION OF PACKAGES AND SERIAL NUMBERS OF PACKAGINGS

4.53. Once packagings have been adequately designed, assessed and manufactured, it is required that they be appropriately identified throughout their lifetime. Paragraphs 531–537 of the Transport Regulations specify the identification marks assigned by the competent authority, the serial numbers of packagings, and the markings of the package types that are required to be present during transport. SSG-26 (Rev. 1) [2] provides recommendations on the legibility, durability and positioning of such markings. In its activities relating to compliance assurance, the competent authority should verify the following:

(a)   All required markings, serial numbers and identification marks are correctly, durably and appropriately applied to packages.

(b)   The user's scheduled inspection and maintenance programme for packagings includes provisions for inspecting and, if necessary, correcting all permanent markings and for repairing any damage or defects.

4.54. The serial number on the packaging is required to uniquely identify each packaging manufactured to a package design approved by the competent authority (see para. 535(b) of the Transport Regulations). For packagings manufactured to an approved Type B(U), Type B(M) or Type C package design and for packagings designed to contain fissile material, it is required that the appropriate competent authority be informed of the serial number (see para. 824 of the Transport Regulations). In this case, the term 'appropriate' has a broad interpretation and could mean any or all of the following:

(a) The competent authority of the State in which the design of the packaging originated;
(b) The competent authority of the State in which the packaging was manufactured;
(c) The competent authority of the State or States in which the packaging is used.

In the case of packagings approved for continued use under para. 820 of the Transport Regulations, all competent authorities involved in the multilateral approval process should be provided with and should retain information on the serial numbers of the packagings.

4.55. An approved package design may be such that different internal components are used with a single outermost component or that the internal components of a packaging are interchangeable between more than one outermost component. In such cases, each outermost component of the packaging with a unique serial number will identify the packaging as an assembly of components; this will satisfy the requirements of para. 535(b) of the Transport Regulations, provided that the assembly of components is in accordance with the design approved by the competent authorities. In such cases, the correct identification and use of the components should be ensured through the management system established by the consignor.

APPROVAL OF SHIPMENTS UNDER SPECIAL ARRANGEMENT

4.56. Paragraph 310 of the Transport Regulations includes provisions for a consignment that does not satisfy all the applicable requirements to be transported under special arrangement, and para. 830 of the Transport Regulations describes the contents of the application for approval of such consignments. For international shipments under special arrangement, multilateral approval is required (see para. 310 of the Transport Regulations). Recommendations on multilateral approval are provided in Section 5.

4.57. For a shipment under special arrangement, the competent authority should verify that the overall level of safety is at least equivalent to that which would be provided if the applicable requirements of the Transport Regulations had been met (see para. 310 of the Transport Regulations). The competent authority should give consideration to the reasons why the shipment cannot be made in full compliance with applicable requirements.

INSPECTION OF TRANSPORT OPERATIONS

4.58. A major feature of the compliance assurance programme of a competent authority will be the inspection of transport operations. Recommendations applicable to such inspections are provided in paras 3.210–3.294 of GSG-13 [4] and address the objectives, organization, types, planning, performance and records of inspections and the follow-up of inspection findings. In addition to producing evidence of compliance, such inspections can be used to verify the degree of compliance by the user as well as the adequacy and suitability of the regulatory requirements. Such inspections may be undertaken during any phase of the transport or during storage in transit and may be announced or unannounced (see paras 3.247–3.249 of GSG-13 [4]). Inspections should, however, be planned sufficiently in advance, and their frequency should be determined in accordance with a graded approach based on the scope of the transport activities of the organization being inspected, and in accordance with the complexity and radiological significance of these activities. Example checklists that could be used for such inspections are provided in Annexes VI and VII.

4.59. Inspections of transport operations should be undertaken by the competent authority or by an organization nominated by it. In some States, such inspections are undertaken on a modal basis by examining all types of dangerous goods; for example, the aviation authority inspects air shipments and the maritime department inspects marine shipments. In such cases, the competent authority may act as an adviser to the organizations that conduct inspections. Users involved in all types and aspects of transport should be periodically inspected, in accordance with a graded approach.

4.60. During inspections of transport operations of users, the competent authority should verify that the following recommendations have been met:

(a) The user's management should provide the necessary personnel and resources to implement an effective programme for compliance with the Transport Regulations. This programme should be part of the management system of the user. The persons responsible for fulfilling specific requirements should be clearly identified. The management system should clearly delegate authority to those responsible persons.

(b) Suitable training should be provided for those persons responsible for implementing the programme for compliance with the Transport Regulations, and this training should be documented.

(c) Packagings appropriate for the contents of the packages should be used.

(d) The user should have all the documentation required by the Transport Regulations, including the relevant certificates of approval by the competent authority and any associated instructions for emergency arrangements; for handling, loading, stowage and use of packages; and for the maintenance of packagings. These instructions are usually in the form of an instruction manual.

(e) The user should follow established procedures for the preparation and use of the packages, in accordance with the certificate of approval, the instruction manual and related documents.

(f) Procedures should be established and followed to properly mark and label packages in accordance with the Transport Regulations. This should include the proper determination and application of the correct transport index.

(g) Procedures should be established and followed, and appropriate and properly calibrated instruments should be provided, to monitor dose rates and contamination levels.

(h) Procedures should be established and followed for the preparation and control of transport documents, for the placarding of vehicles, for the provision of documentation for carriers and for the notification of competent authorities.

(i) During transport, carriers should perform the required actions for placarding and for the stowage and separation of packages in accordance with the Transport Regulations. Carriers should also undertake any administrative controls relating to exclusive use shipments or any supplementary operational controls specified in the certificate of approval of the competent authority.

(j) Procedures should be established to respond to cases of non-compliance in order to meet the requirements of para. 309 of the Transport Regulations.

More detailed recommendations on the corresponding inspection activities of the competent authority are provided in paras 4.67–4.82.

4.61. When preparing for inspections of transport operations, the competent authority should consider those requirements that apply to industrial packages in the Transport Regulations and those requirements that originate from the provisions of international conventions and standards (see Refs [12, 32]).

4.62. The requirements for notification of the competent authority regarding certain packages and shipments are established in paras 557–560 of the Transport Regulations. The competent authority may request additional notification before a package is shipped, or after it has been received, so that plans can be made for certain inspections. This need for additional notification should be determined in accordance with the package types and the number of shipments made and received. Furthermore, if users are required by national regulations to submit reports to the competent authority concerning the transport of radioactive material, the information in these reports should be used by the competent authority to assess the status of the transport of radioactive material within the country and should be considered in establishing the nature and extent of its activities related to compliance assurance.

4.63. Recommendations on inspection performance and follow-up actions are provided in GSG-13 [4].

RADIATION PROTECTION

4.64. Paragraphs 301–303 of the Transport Regulations establish the general requirements for radiation protection and the requirements for radiation protection programmes in the transport of radioactive material. Through its compliance assurance programme, the competent authority should ensure that these requirements have been met, for example by requesting information on and inspecting the radiation protection programmes for transport. Specific recommendations on radiation protection programmes for the safe transport of radioactive material are provided in IAEA Safety Standards Series No. TS-G-1.3, Radiation Protection Programmes for the Transport of Radioactive Material [33], including items to be reviewed to evaluate the contents and effectiveness of radiation protection programmes.

4.65. When necessary, the competent authority should require the inclusion of information on radiation protection programmes in applications for approval of shipments or special arrangements.

4.66. The competent authority is required to arrange for periodic assessments to evaluate the radiation doses to workers and to members of the public due to the transport of radioactive material (see para. 308 of the Transport Regulations). Data from consignors and carriers that need to assess the doses arising from their transport operations may be used in such assessments of radiation doses by the competent authority. However, the competent authority should independently verify the data received from consignors and carriers. Questionnaires, analyses, site visits and measurements may be used to assess doses.

INSPECTION OF MANUFACTURING

4.67. During the manufacture of special form radioactive material or low dispersible radioactive material, the competent authority should, as part of the compliance assurance programme, conduct inspections of the management system of the manufacturer (see Annexes IV and V) and inspections of the manufacturing operations to ensure that all the requirements have been correctly implemented.

4.68. The competent authority should give particular consideration to how the management system is applied before the manufacture of packagings begins, such as during the development of manufacturing processes and procedures. The inspection programme for the manufacture of a single packaging may be different from that for the continuous manufacture of packagings.

4.69. Packagings should be manufactured in accordance with the design specifications through a process that is subject to the management system. To confirm this, the competent authority should perform inspections of the manufacturing process, including the actual implementation and effectiveness of the management system. The management system may be inspected by the competent authority before the commencement of manufacturing of a packaging and periodically thereafter (see also para. 4.23(b) and (c)). An example checklist for inspections of the manufacturing of packagings is provided in Annex VIII.

4.70. Facilities operated by manufacturers and their subcontractors may be inspected by the competent authority. The frequency and extent of such inspections should be determined in accordance with a graded approach based on the confidence the competent authority has in the manufacturing arrangements and the importance to safety of the items being manufactured.

4.71. The inspection process may include taking samples for independent non-destructive or destructive testing. The purpose is to verify that the packaging is manufactured in compliance with the Transport Regulations and in accordance with the design specification.

4.72. Reports of deviations from specifications and reports of repairs that have been performed should be made available to the competent authority for review.

The competent authority should have the authority to accept or reject any deviations from the approved specifications.

4.73. The results of inspections by the competent authority should be recorded and communicated to the manufacturer (see GSG-13 [4]) and other responsible parties — for example, the packaging owner or the packaging designer — for information and for possible action.

4.74. During inspections of manufacturers, compliance with the requirements of para. 501 of the Transport Regulations regarding the first use of a packaging should be verified. From the results of quality control tests, reports on deviations, and other measures within the management system, the manufacturer should verify that the packaging has been manufactured in compliance with the specification issued by the package designer in accordance with the package design safety report and the certificate of approval, if a certificate has been issued.

INSPECTION OF MAINTENANCE OPERATIONS

4.75. The competent authority should verify that maintenance operations, as specified by the original designer or the competent authority, have been conducted by a person or organization that has an appropriate management system (see para. 306 of the Transport Regulations). In particular, any proposed modifications to a packaging during maintenance operations should be implemented only when the necessary modification specifications are available to the person or organization implementing the modification. Any departure from these specifications might render the packaging unusable and might compromise the original design intent. An example checklist for inspections of maintenance operations is provided in Annex IX.

4.76. The competent authority should verify that appropriate records of all maintenance operations demonstrate that the package fully complies with the requirements specified in the certificate of approval and the relevant requirements of the Transport Regulations.

4.77. Inspections by the competent authority should include the packaging, the storage locations for both the packaging and records, the packaging maintenance facility and any other factors that could affect the lifetime of the packaging. For packages intended to be used for shipment after storage, special attention should be given to the correct implementation of the maintenance instructions that contribute to ageing management (para. 503(e) of the Transport Regulations).

Inspections should include verification of the appropriate use of records and logbooks (as described in TS-G-1.4 [28]) if maintenance operations are performed at different locations.

4.78. The user should record all safety related deviations from the design specifications, as well as any significant damage noted during the use of the packages. Reports of deviations from specifications and of repairs that have been performed should be made available to the competent authority for review in accordance with the requirements of the competent authority. The competent authority should have the authority to accept or reject any deviations from the approved specifications. Corrective measures or modification proposals, including any plans for repairs, might be subject to the agreement of the competent authority. In such cases, any packages undergoing repairs, modifications or changes should not be returned to use until the competent authority has agreed to or approved the change.

INSPECTION OF CONSIGNORS

4.79. The consignor may be the owner of the package, the manufacturer of the package, or the user or operator of a package owned by a third party. The consignor may delegate some of the actions needed to prepare a package for transport in accordance with the Transport Regulations, but the consignor retains overall responsibility for these actions. The declaration on the transport documents signed by the consignor attests to this responsibility.

4.80. The competent authority should ensure that the consignor's responsibilities, as defined in paras 545–561 of the Transport Regulations, are followed (see also para. 4.23(d)). An example checklist that could be used for inspections of consignors is provided in Annex VI. The competent authority should inspect compliance with the following, as appropriate:

(a)   The consignor should have an appropriate and functioning management system to cover all aspects of its responsibilities and activities in the transport of radioactive material. If a consignor consigns only one type of package infrequently, the consignor may control and conduct all activities directly. A consignor that produces or reuses a large number of different package types may use different contractors for different parts of the work, but the activities of such contractors should be provided for and controlled by means of the consignor's management system.

(b)   The consignor should have a clear understanding of the nature, form and activity of the radioactive material to be consigned.

(c)   The consignor should fill or load the material into the packaging for transport in accordance with the package requirements and instructions. This could involve, for example, verifying that the contents of the package have been positioned correctly within the packaging to maximize the shielding protection afforded by the packaging.

(d)   The consignor should use appropriate packaging for which there is a valid certificate of approval or appropriate documentary evidence of compliance. Package design approvals should be valid for the entire duration of the journey and should not expire in the course of long international transport. Also, the certificate of approval is required to cover the entire radioactive contents permitted to be carried, and the consignor is required to have the correct certificate for the contents being transported in accordance with the Transport Regulations.

(e)   The consignor should have the relevant packing instructions for the package; copies of these instructions are required to be available at the location where the package is prepared for transport. The packing instructions provide detailed information and instructions on the loading configuration of the contents, the closure methods and the tightening torques of fasteners; to be followed by the consignor.

(f)   The consignor should ensure that the package used for transport conforms to its specifications, including those indicated on a certificate of approval for a package design that requires competent authority approval, and that the packaging is in an acceptable condition. For packagings, the consignor should have evidence, such as certificates of conformity or inspection reports, that indicate that the packagings conform to their specifications, including those indicated on a certificate of approval for a package design that requires competent authority approval. In the case of reusable packagings, the consignor should have evidence (e.g. in the form of inspection reports, release notes and certificates of conformity) that all necessary and specified maintenance work has been performed and that the packaging is suitable for the next complete transport operation or programme of movements. The consignor should not use a package that does not comply with the approved specifications or that has not been subjected to the required maintenance.

(g)   The consignor should complete and apply the correct labels and markings for packages when they are presented for transport. For example, the consignor should determine the transport index and should have correctly functioning and calibrated monitoring instruments for measuring the dose rates of the package, the overpack, the freight container and the vehicle.

(h) The consignor should have appropriate, calibrated monitoring instruments and trained staff so that the staff can measure dose rates and radioactive contamination associated with the transport of radioactive material. For example, the consignor should be able to satisfy the competent authority that its staff is knowledgeable in the operation of the monitoring instruments and is capable of conducting correct measurements of dose rates and radioactive contamination as required by the Transport Regulations.

(i) Subject to national legislation, the consignor should have the necessary licences or other permissions, granted by the competent authority or by other governmental bodies, to function as a consignor of radioactive material. Also, the competent authority should be satisfied that the consignor has the approvals required for the transport of the radioactive material (e.g. shipment approval, special form approval). Multiple approvals may be necessary for the transport of radioactive material where considerations of security and considerations of nuclear material accounting and control apply.

(j) The consignor is required to complete the required transport documents, giving the appropriate information, as specified in paras 546–555 of the Transport Regulations. The consignor should also provide the transport documents to the carrier to enable the carrier, or any subsequent carriers, to meet any other applicable national or international modal regulations. The consignor is required to retain a copy of each transport document for a minimum period of three months (see para. 555 of the Transport Regulations). During inspections by the competent authority, it should be verified that complete and accurate information is given in the transport documents (sometimes called 'shipper's certificates' or 'consignment notes'). It should be verified that the transport documents take account of any variations imposed by national or international modal regulations. It should also be verified that the transport documents cover the entire journey of the consignment.

(k) The consignor is required to provide information to the carriers, in accordance with para. 554 of the Transport Regulations. The competent authority should verify that the required information and documents have been provided to the carriers by inspecting both the consignor and the carriers (see para. 4.81).

(l) The consignor should notify the competent authorities of transport movements, as required by paras 557–560 and summarized in annex I to the Transport Regulations. Through its inspections of consignors and its liaison with other competent authorities, the responsible competent authority should verify that the required notifications are being made.

(m) The consignor is required, before each shipment of any package, to ensure that the requirements specified in the relevant provisions of the Transport

Regulations and in the applicable certificates of approval have been fulfilled (see para. 503 of the Transport Regulations). During the lifetime of a packaging, the consignor should maintain records demonstrating that the requirements of para. 503 of the Transport Regulations have been met. The competent authority should verify that the consignor's management system provides controls to ensure that all required pre-dispatch activities have been specified and completed and that the declaration and signature of the final consignor are valid.

(n)   The consignor should have appropriate procedures in place to detect cases of non-compliance and to respond in accordance with the requirements of para. 309 of the Transport Regulations.

(o)   The activities of the consignor are covered by a radiation protection programme that meets the requirements of para. 302 of the Transport Regulations. Further recommendations on radiation protection programmes are provided in paras 4.64 and 4.65 of this Safety Guide.

INSPECTION OF CARRIERS

4.81. The competent authority should conduct inspections (see para. 4.59) to verify that the carrier is contributing to the safe transport of radioactive material as follows:

(a)   The carrier should have an appropriate management system that meets the requirements of para. 306 of the Transport Regulations and covers all relevant aspects of the carrier's responsibilities and activities in the transport of radioactive material. A carrier that occasionally carries one type of package within national boundaries using one mode of transport may have a relatively simple management system. In contrast, a national or international carrier that frequently carries large numbers of packages and operates a multimodal carriage and distribution service will need a more comprehensive management system to control its activities and to ensure compliance with the Transport Regulations.

(b)   The carrier should have sufficient knowledge of national and international regulations to understand the information and documents provided by the consignor. The carrier should have knowledge and understanding of the requirements for transport documents established in paras 546–554 of the Transport Regulations and implement procedures for checking the validity and accuracy of such documents.

(c)   The carrier should have knowledge of, and the ability and resources to meet, additional provisions concerning loading, stowage, transport, handling and unloading of packages, as well as the ability to comply with any restrictions on routeing, means of conveyance or mode of transport. For conveyances such as trucks or railway wagons, the carrier should have the necessary facilities or equipment for achieving secure tie-down arrangements and should comply with any additional speed limits that are specified. Also, if escort vehicles and personnel are required for the transport operations, the carrier should demonstrate to the competent authority that it can provide them.

(d)   The carrier should be able to identify damaged or poorly prepared packages. The carrier should be familiar with the required placards, package labels and markings, should understand their meaning and purpose, and should be able to relate the information displayed to the details given in the transport documents. The carrier should have the appropriate procedures and the necessary understanding to ensure that any damaged, poorly prepared or incorrectly labelled packages are rejected or quarantined, that packages are correctly stowed within the vehicle and that basic checks of the transport documents against the package labels will be conducted.

(e)   The carrier should operate vehicles or other means of conveyance that can be used to carry the radioactive material or packages safely, without overloading, without infringing the required segregation distances and without exceeding the limitations on the transport index and the criticality safety index. The carrier should ensure that, when required, the number, type, size and location of placards on the conveyance meet the regulatory requirements.

(f)   The carrier is required to establish appropriate arrangements for a nuclear or radiological emergency (see paras 304 and 305 of the Transport Regulations). These arrangements should take account of the types of radioactive material being carried and the type of conveyance being used. The carrier may have its own emergency arrangements; alternatively, the carrier may participate in or use the consignor's emergency arrangements or other national emergency schemes or arrangements. Irrespective of the emergency arrangements that apply, the carrier should be familiar with the arrangements in place, and all personnel involved should receive the necessary training in the emergency arrangements.

(g)   The carrier should have the capability to implement appropriate controls in connection with storage in transit, in particular with regard to the safety of workers and the public. The carrier is required to implement the provisions established in paras 562 and 563 of the Transport Regulations concerning the segregation of packages during transport and storage in transit.

(h)   The carrier should have appropriate procedures in place to identify cases of non-compliance and to take appropriate action in accordance with the requirements of para. 309 of the Transport Regulations.

(i)   The activities of the carrier are covered by a radiation protection programme that meets the requirements of para. 302 of the Transport Regulations. Further recommendations on radiation protection programmes are provided in paras 4.64 and 4.65 of this Safety Guide.

An example checklist that could be used for inspections of carriers is provided in Annex VII.

INSPECTION OF CONSIGNEES

4.82. During inspection of the transport operations of the consignee, the responsible competent authority should verify that the following recommendations are observed:

(a)   The consignee should have a management system that meets the requirements of para. 306 of the Transport Regulations and that covers all applicable activities.

(b)   A radiation protection programme that meets the requirements of para. 302 of the Transport Regulations should be established and implemented. Further recommendations on radiation protection programmes are provided in paras 4.64 and 4.65 of this Safety Guide.

(c)   Workers of the consignee should receive appropriate training commensurate with their duties.

(d)   The consignee should be able to take appropriate actions in cases of non-compliance, in accordance with the requirements of para. 309 of the Transport Regulations.

CONTENTS OF PACKAGES WITH OTHER DANGEROUS PROPERTIES

4.83. In addition to the radioactive and fissile properties, any other dangerous properties of the contents of the package, such as chemical toxicity and corrosiveness, are required to be addressed through compliance with the relevant transport regulations for dangerous goods (see para. 507 of the Transport Regulations). This may involve liaison and cooperation between the competent authority and other governmental bodies that have a responsibility in such matters.

EMERGENCY PREPAREDNESS AND RESPONSE

4.84. Activities related to emergency preparedness and response are among the fundamental activities of the competent authority (see para. 4.4). Requirements for emergency preparedness and response are established in IAEA Safety Standards Series No. GSR Part 7, Preparedness and Response for a Nuclear or Radiological Emergency [34]. The competent authority has several relevant roles and responsibilities with regard to emergency preparedness and response, including reviewing the arrangements for emergency preparedness and response of users during inspections (see para. 4.81(f) and Annexes III, VI and VII) and during the review of applications for approvals; issuing approvals (Annex I); establishing roles and responsibilities; liaising with other relevant governmental agencies (see paras 2.17–2.19 and 2.22); participating in exercises; participating in training; and maintaining appropriate expertise (see para. 2.15). Detailed recommendations on emergency preparedness and response in the transport of radioactive material are provided in SSG-65 [9].

4.85. International cooperation might be necessary when States are affected by accidents that occur during the transport of radioactive material. Certain types of transport accident are covered by the Convention on Early Notification of a Nuclear Accident and the Convention on Assistance in the Case of a Nuclear Accident or Radiological Emergency [35].

ENFORCEMENT ACTIONS AND INVESTIGATIONS OF INCIDENTS

4.86. The compliance assurance programme should include provisions for enforcement consistent with the requirements established in GSR Part 1 (Rev. 1) [6] and with the recommendations provided in GSG-13 [4].

4.87. Paragraph 2.5 of GSR Part 1 (Rev. 1) [6] requires that the government promulgate laws and statutes to make provision for an effective governmental, legal and regulatory framework for safety, including provision for the enforcement of regulations, in accordance with a graded approach.

4.88. Requirement 30 of GSR Part 1 (Rev. 1) [6] states:

> **"The regulatory body shall establish and implement an enforcement policy within the legal framework for responding to non-compliance by authorized parties with regulatory requirements or with any conditions specified in the authorization."**

4.89. Paragraph 4.55 of GSR Part 1 (Rev. 1) [6] states:

> "Enforcement actions by the regulatory body may include recorded verbal notification, written notification, imposition of additional regulatory requirements and conditions, written warnings, penalties and, ultimately, revocation of the authorization. Regulatory enforcement may also entail prosecution, especially in cases where the authorized party does not cooperate satisfactorily in the remediation or resolution of the non-compliance."

4.90. Recommendations on the objectives of enforcement, methods of enforcement, factors in determining enforcement actions, the inspector's authority in relation to enforcement, use of the enforcement process and records of enforcement are provided in GSG-13 [4].

4.91. The enforcement activities of the competent authority should be applicable to all activities important to safety in transport, irrespective of whether a certificate of approval from the competent authority is required.

4.92. Users should be required by national regulations to report to the competent authority all significant incidents, including accidents or significant non-compliance with the Transport Regulations. The competent authority should investigate any reported incidents, in accordance with a graded approach. Such investigations may include gathering information through special inspections and/or during routine inspections.

4.93. In para. 2.18, it is recommended that the competent authority facilitate cooperation between all governmental bodies involved in the transport of radioactive material. One of the aims of such cooperation is to ensure the consistent application of enforcement measures relating to compliance assurance.

4.94. Recognizing the international aspect of transport, incident investigation and enforcement may necessitate international cooperation between States.

MAINTENANCE OF REGULATIONS AND FEEDBACK TO THE COMPETENT AUTHORITY

4.95. Recommendations for the development, review and revision of regulations and guides are provided in paras 3.51–3.71 of GSG-13 [4]. Consistent with these recommendations, the competent authority should periodically review national and international regulations for the transport of radioactive material and should

make any necessary changes to the national regulations. The competent authority
should maintain awareness of developments in international organizations
(e.g. the International Maritime Organization, the International Civil Aviation
Organization) and conventions, and of any associated mandatory timescales for
changes to be introduced (see also Section 3).

5. MULTILATERAL APPROVALS

5.1.   Under the Transport Regulations, multilateral approval of the items listed in
para. 4.12 may be effected in either of the following ways:

(a)   Independent approval by the responsible competent authority of each
      country through or into which the consignment is to be transported (i.e. a
      chain of multilateral competent authority approvals);
(b)   Validation of the original certificate issued by the competent authority of the
      country of origin of the design or shipment in accordance with para. 840 of
      the Transport Regulations.

Both an independent approval and a validation may cover either all parts of the
original certificate requiring multilateral approval (full multilateral approval) or
only the parts deemed appropriate by the applicant or the competent authority
(partial multilateral approval). The issued certificate of approval or validation
of each competent authority involved in such a multilateral approval process is
only applicable within its territory of jurisdiction (see also para. 6.4). It is the
responsibility of the consignor to receive all applicable multilateral approvals for
each country through or into which its consignment is to be transported before
the transport takes place.

5.2.   The competent authority should communicate to the applicants its policy
on how multilateral approval is performed (i.e. guidelines on which type of
approval — independent approval or validation — will be issued for which type
of design or shipment). The competent authority's policy may be based on criteria
such as the risk associated with the use of the package in its territory.

5.3.   Independent approval provides more flexibility in determining the extent
of the multilateral approval. This is useful if modification of any of the essential
detailed provisions of the certificate of approval of the original competent authority
is deemed necessary or if new provisions are to be added to the approval. In such

cases, an independent assessment of the application should be performed by the relevant competent authority.

5.4. The essential difference between an independent approval and a validation is that the latter is not self-contained; that is, some reference to the original certificate of approval is made (e.g. for description of the packaging or contents or for shipment provisions). For convenience to local users, the validation may, however, contain parts or summaries of parts of the original certificate of approval, in translation if necessary.

5.5. A validation generally reduces, but does not necessarily exclude, the possibility of differences between provisions of certificates issued by different competent authorities that cover the same case. Such differences might arise because of supplementary or divergent local regulations or different practices of the competent authorities.

5.6. An endorsement is a special kind of validation that simply states that all provisions of the original certificate of approval are endorsed. An endorsement may contain supplementary provisions or information if they do not conflict with the provisions of the original certificate and do not modify the design. An endorsement should use the identification mark of the original certificate (i.e. not a separate identification mark).

5.7. The competent authority of any country that the shipment is to be transported through or into should take part in any chain of multilateral approvals by competent authorities. The competent authorities of the countries from which a vessel or aircraft departs and at which it arrives, as well as the competent authority of the flag state of the vessel or aircraft (which is considered to be part of the territory of the flag State), may be involved in the multilateral approval process.

5.8. Multilateral approvals should not be issued before the certificate of approval is issued by the competent authority of the country of origin of the design or shipment. However, when a competent authority is requested to give its approval as part of a chain of multilateral approvals, parallel assessment of the application may be considered at the discretion of the competent authority.

6. INTERNATIONAL COOPERATION BETWEEN COMPETENT AUTHORITIES CONCERNING PACKAGES AND SHIPMENTS OF FOREIGN ORIGIN

INTERNATIONAL COOPERATION RELATING TO COMPLIANCE ASSURANCE

6.1. The national competent authority is responsible for compliance assurance within its territory. However, many shipments of radioactive material involve packages of foreign origin. Each such instance of transport should also comply with national regulatory requirements (see also paras 3.5 and 3.8).

6.2. To ensure compliance with the Transport Regulations in the case of the transport of radioactive material of foreign origin transiting its area of jurisdiction, the competent authority should consider inspecting such packages or shipments. Cooperation with other national competent authorities should also be considered.

6.3. National competent authorities should cooperate to further develop the Transport Regulations and its associated advisory and explanatory Safety Guides for the safe transport of radioactive material [2, 9, 28, 31, 33, 36]. One objective of such cooperation is the uniform application of the requirements of the Transport Regulations in all Member States.

PACKAGES AND SHIPMENTS OF FOREIGN ORIGIN THAT ARE SUBJECT TO MULTILATERAL APPROVAL

6.4. Foreign packages and shipments listed in para. 4.12 require multilateral approval by the competent authority of each country through or into which the consignment is to be transported. For such cases, the competent authority that is requested to issue a validation of the original certificate of approval can request detailed information relating to the design assessment and the management system before it issues a certificate of validation. Cooperation between the validating competent authority and the competent authority that issued the original certificate of approval will contribute to ensuring that the necessary compliance assurance is provided.

6.5. In cases where there is doubt about a specific management system, the validating competent authority should contact the competent authority of the

State of origin of the package or shipment and should request relevant details of inspections. Where important shipments or large scale operations are concerned, efforts to ensure international cooperation may involve visits between competent authorities and joint visits of the respective organizations for detailed discussion of the management system. The purpose of such visits is to gain confidence in the standards used in different States and to reach agreement on the approach concerning differences in standards.

6.6. Where multilateral approval is effected by the issuance of independent certificates by successive countries, the competent authority should verify that the validating mark, as required by para. 833(b) of the Transport Regulations, is legibly and durably marked on the packaging.

6.7. For operations associated with foreign packages and shipments, notifications to the competent authority of each country through or into which the consignment is to be transported are also required, in accordance with paras 557–559 of the Transport Regulations.

PACKAGES AND SHIPMENTS OF FOREIGN ORIGIN THAT DO NOT REQUIRE NOTIFICATION OF THE COMPETENT AUTHORITY

6.8. Transport of radioactive material that does not require the competent authority to be notified, particularly packages and shipments of foreign origin, may nevertheless be subject to inspections by the competent authority. International cooperation between competent authorities can be used to inform interested parties about such transport, but competent authorities may also identify such packages and shipments in the same way as the transport of other dangerous goods.

6.9. The competent authority might receive only the notifications required under paras 557 and 558 of the Transport Regulations. Nevertheless, the carrier will be in possession of the transport documents supplied by the consignor, which will contain the information required in paras 546–554 of the Transport Regulations. The competent authority should check this information as part of its compliance assurance programme.

6.10. In some States, information arising from the legal requirements associated with, for example, the shipment of certain radioactive material across national borders, or the legal requirements arising from international protocols and/or codes of conduct established to facilitate cooperation between competent authorities, might also be used to augment any information obtained from consignors or carriers.

REFERENCES

[1]   INTERNATIONAL ATOMIC ENERGY AGENCY, Regulations for the Safe Transport of Radioactive Material, 2018 Edition, IAEA Safety Standards Series No. SSR-6 (Rev. 1), IAEA, Vienna (2018).

[2]   INTERNATIONAL ATOMIC ENERGY AGENCY, Advisory Material for the IAEA Regulations for the Safe Transport of Radioactive Material (2018 Edition), IAEA Safety Standards Series No. SSR-26 (Rev. 1), IAEA, Vienna (2022).

[3]   INTERNATIONAL ATOMIC ENERGY AGENCY, Organization, Management and Staffing of the Regulatory Body for Safety, IAEA Safety Standards Series No. GSG-12, IAEA, Vienna (2018).

[4]   INTERNATIONAL ATOMIC ENERGY AGENCY, Functions and Processes of the Regulatory Body for Safety, IAEA Safety Standards Series No. GSG-13, IAEA, Vienna (2018).

[5]   INTERNATIONAL ATOMIC ENERGY AGENCY, Leadership and Management for Safety, IAEA Safety Standards Series No. GSR Part 2, IAEA, Vienna (2016).

[6]   INTERNATIONAL ATOMIC ENERGY AGENCY, Governmental, Legal and Regulatory Framework for Safety, IAEA Safety Standards Series No. GSR Part 1 (Rev. 1), IAEA, Vienna (2016).

[7]   INTERNATIONAL ATOMIC ENERGY AGENCY, Establishing the Infrastructure for Radiation Safety, IAEA Safety Standards Series No. SSG-44, IAEA, Vienna (2018).

[8]   INTERNATIONAL ATOMIC ENERGY AGENCY, IAEA Nuclear Safety and Security Glossary: Terminology Used in Nuclear Safety, Nuclear Security, Radiation Protection and Emergency Preparedness and Response, 2022 (Interim) Edition, IAEA, Vienna (2022).

[9]   INTERNATIONAL ATOMIC ENERGY AGENCY, INTERNATIONAL CIVIL AVIATION ORGANIZATION, INTERNATIONAL MARITIME ORGANIZATION, Preparedness and Response for a Nuclear or Radiological Emergency Involving the Transport of Radioactive Material, IAEA Safety Standards Series No. SSG-65, IAEA, Vienna (2022).

[10]   The Convention on the Physical Protection of Nuclear Material, INFCIRC/274/Rev. 1, IAEA, Vienna (1980).

[11]   Amendment to the Convention on the Physical Protection of Nuclear Material, INFCIRC/274/Rev. 1/Mod. 1, IAEA, Vienna (2016).

[12]   UNITED NATIONS, Recommendations on the Transport of Dangerous Goods, Model Regulations, ST/SG/AC.10/1/Rev.22, 2 vols, United Nations, New York and Geneva (2019).

[13] INTERNATIONAL CIVIL AVIATION ORGANIZATION, Technical Instructions for the Safe Transport of Dangerous Goods by Air, 2021–2022 Edition, ICAO, Montreal (2020).

[14] INTERNATIONAL MARITIME ORGANIZATION, International Maritime Dangerous Goods Code (IMDG Code), Amendment 40-20, 2020 Edition, IMO, London (2020).

[15] INTERNATIONAL ATOMIC ENERGY AGENCY, Nuclear Security Recommendations on Physical Protection of Nuclear Material and Nuclear Facilities (INFCIRC/225/Revision 5), IAEA Nuclear Security Series No. 13, IAEA, Vienna (2011).

[16] INTERNATIONAL ATOMIC ENERGY AGENCY, Nuclear Security Recommendations on Radioactive Material and Associated Facilities, IAEA Nuclear Security Series No. 14, IAEA, Vienna (2011).

[17] INTERNATIONAL ATOMIC ENERGY AGENCY, Security of Nuclear Material in Transport, IAEA Nuclear Security Series No. 26-G, IAEA, Vienna (2015).

[18] INTERNATIONAL ATOMIC ENERGY AGENCY, Security of Radioactive Material in Transport, IAEA Nuclear Security Series No. 9-G (Rev. 1), IAEA, Vienna (2020).

[19] INTERNATIONAL CIVIL AVIATION ORGANIZATION, Convention on International Civil Aviation (Chicago Convention), Doc 7300/9, Ninth Edition, ICAO, Montreal (2006).

[20] INTERNATIONAL MARITIME ORGANIZATION, International Convention for the Safety of Life at Sea (SOLAS), 1974, as amended, IMO, London (1974).

[21] UNIVERSAL POSTAL UNION, Convention Manual, Update 2 — October 2020, UPU, Berne (2020).

[22] UNITED NATIONS ECONOMIC COMMISSION FOR EUROPE, INLAND TRANSPORT COMMITTEE, Agreement Concerning the International Carriage of Dangerous Goods by Road, UNECE, New York and Geneva (2020).

[23] INTERGOVERNMENTAL ORGANISATION FOR INTERNATIONAL CARRIAGE BY RAIL, Convention Concerning International Carriage by Rail, OTIF, Berne (1999).

[24] INTERGOVERNMENTAL ORGANISATION FOR INTERNATIONAL CARRIAGE BY RAIL, Regulations concerning the International Carriage of Dangerous Goods by Rail, 2021 Edition, OTIF, Berne (2020).

[25] UNITED NATIONS ECONOMIC COMMISSION FOR EUROPE, COMMITTEE ON INLAND TRANSPORT, European Agreement Concerning the International Carriage of Dangerous Goods by Inland Waterways, UNECE, New York and Geneva (2020).

[26] The MERCOSUR/MERCOSUL Agreement of Partial Reach to Facilitate the Transport of Dangerous Goods (1994).

[27] ORGANIZATION FOR COOPERATION OF RAILWAYS, Agreement on International Goods Traffic by Rail, Warsaw (1951).

[28] INTERNATIONAL ATOMIC ENERGY AGENCY, The Management System for the Safe Transport of Radioactive Material, IAEA Safety Standards Series No. TS-G-1.4, IAEA, Vienna (2008).

[29] INTERNATIONAL ATOMIC ENERGY AGENCY, Safe Transport of Radioactive Material, Fourth Edition, Training Course Series No. 1, IAEA, Vienna (2006).

[30] INTERNATIONAL ATOMIC ENERGY AGENCY, Communication and Consultation with Interested Parties by the Regulatory Body, IAEA Safety Standards Series No. GSG-6, IAEA, Vienna (2017).

[31] INTERNATIONAL ATOMIC ENERGY AGENCY, Format and Content of the Package Design Safety Report for the Transport of Radioactive Material, IAEA Safety Standards Series No. SSG-66, IAEA, Vienna (2022).

[32] INTERNATIONAL ORGANIZATION FOR STANDARDIZATION, Series 1 Freight Containers — Specification and Testing — Part 1: General Cargo Containers for General Purposes, ISO 1496 1:1990, ISO, Geneva (1990); Amendments 1:1993, 2:1998, 3:2005, 4:2006, 5:2006 and ISO 1496 1:2013.

[33] INTERNATIONAL ATOMIC ENERGY AGENCY, Radiation Protection Programmes for the Transport of Radioactive Material, IAEA Safety Standards Series No. TS-G-1.3, IAEA, Vienna (2007). (A revision of this publication is in preparation.)

[34] FOOD AND AGRICULTURE ORGANIZATION OF THE UNITED NATIONS, INTERNATIONAL ATOMIC ENERGY AGENCY, INTERNATIONAL CIVIL AVIATION ORGANIZATION, INTERNATIONAL LABOUR ORGANIZATION, INTERNATIONAL MARITIME ORGANIZATION, INTERPOL, OECD NUCLEAR ENERGY AGENCY, PAN AMERICAN HEALTH ORGANIZATION, PREPARATORY COMMISSION FOR THE COMPREHENSIVE NUCLEAR TEST BAN TREATY ORGANIZATION, UNITED NATIONS ENVIRONMENT PROGRAMME, UNITED NATIONS OFFICE FOR THE COORDINATION OF HUMANITARIAN AFFAIRS, WORLD HEALTH ORGANIZATION, WORLD METEOROLOGICAL ORGANIZATION, Preparedness and Response for a Nuclear or Radiological Emergency, IAEA Safety Standards Series No. GSR Part 7, IAEA, Vienna (2015).

[35] INTERNATIONAL ATOMIC ENERGY AGENCY, Convention on Early Notification of a Nuclear Accident and Convention on Assistance in the Case of a Nuclear Accident or Radiological Emergency, Legal Series No. 14, IAEA, Vienna (1987).

[36] INTERNATIONAL ATOMIC ENERGY AGENCY, Schedules of Provisions of the IAEA Regulations for the Safe Transport of Radioactive Material (2018 Edition), IAEA Safety Standards Series No. SSG-33 (Rev. 1), IAEA, Vienna (2021).

NOTE ON THE ANNEXES

The annexes provide example templates, procedures and checklists that may be used by a competent authority in performing various functions and activities that are part of a compliance assurance programme. If this material is used by a Member State, it will need to be adapted in accordance with national regulatory requirements, working practices and methods. The material in the annexes has not been endorsed by the IAEA or its Member States.

The information in the section titled 'Information on Management Systems' in Annex I has been adapted from US Nuclear Regulatory Commission Regulatory Guide 7.10, Revision 2, Establishing Quality Assurance Programs for Packaging Used in Transport of Radioactive Material[1]. The checklists in Annexes VI–IX have been adapted from the European Association of Competent Authorities' Technical Guide: Compliance Inspections by the European Competent Authorities on the Transport of Radioactive Material, Issue 1[2].

---

[1] NUCLEAR REGULATORY COMMISSION, Establishing Quality Assurance Programs for Packaging Used in Transport of Radioactive Material, Regulatory Guide 7.10, Revision 2, Office of Nuclear Regulatory Research, Washington, DC (2005).

[2] EUROPEAN ASSOCIATION OF COMPETENT AUTHORITIES, Technical Guide: Compliance Inspections by the European Competent Authorities on the Transport of Radioactive Material, Issue 1, EACA (2015).

Annex I

INFORMATION TO BE INCLUDED IN
APPLICATIONS FOR APPROVALS

I–1.   This annex provides details of the information to be included in applications for approval of the following:

(a)   Design of packages;
(b)   Design of special form radioactive material and low dispersible radioactive material;
(c)   Shipments;
(d)   Shipments under special arrangement;
(e)   Management systems.

INFORMATION TO BE INCLUDED IN APPLICATIONS FOR APPROVAL OF DESIGN OF PACKAGES

I–2.   The applicant seeking approval needs to provide the competent authority with all necessary information to demonstrate that the package design meets all applicable regulatory requirements. The corresponding application document, which is sometimes referred to as the package design safety report, needs to at least contain the information as specified in paras 807(c), 809, 812 and 815 of IAEA Safety Standards Series No. SSR-6 (Rev. 1), Regulations for the Safe Transport of Radioactive Material, 2018 Edition [I–1] (hereinafter referred to as the 'Transport Regulations').

I–3.   Specific recommendations on the formatting and contents of a package design safety report are provided in IAEA Safety Standards Series No. SSG-66, Format and Content of the Package Design Safety Report for the Transport of Radioactive Material [I–2]. SSG-66 [I–2] covers all types of package and assists in the preparation of the package design safety report to demonstrate compliance of a design of a package with all applicable requirements of the Transport Regulations. It provides detailed guidance on the structure and the contents of a package design safety report. It includes all package designs requiring competent authority approval (i.e. Type B(U), Type B(M), Type C, packages containing fissile material and packages designed to contain 0.1 kg or more of uranium hexafluoride). In addition, recommendations are provided for package designs not requiring competent authority approval (i.e. excepted packages, industrial packages (i.e. Type IP-1, Type IP-2, Type IP-3), and Type A packages) to demonstrate compliance with all applicable requirements of the Transport Regulations.

INFORMATION TO BE INCLUDED IN APPLICATIONS FOR
APPROVAL OF DESIGN OF SPECIAL FORM RADIOACTIVE
MATERIAL AND LOW DISPERSIBLE RADIOACTIVE MATERIAL

I–4. The following information is to be included in applications for approval of design of special form radioactive material and low dispersible radioactive material:

(a) General information:
  (i) General description of the design and of the intended use of the special form or low dispersible radioactive material.
  (ii) List of applicable national and international regulations and the edition of the Transport Regulations under which competent authority approval is sought.
(b) Administrative information, including the following:
  (i) Name, address, telephone number and email address of the applicant;
  (ii) Name, address, telephone number and email address of the designer;
  (iii) Type of approval required (i.e. special form or low dispersible radioactive material);
  (iv) Identification mark of the competent authority, if previously allocated;
  (v) General arrangement drawing number;
  (vi) Date of application;
  (vii) Date by which approval is desired.
(c) Specific information as required by para. 803 of the Transport Regulations. A detailed description of the radioactive material or, if a capsule, the contents, including the following:
  (i) Radionuclides present;
  (ii) Total activity;
  (iii) Nature of emitted radiation;
  (iv) Heat output;
  (v) Physical and chemical state;
  (vi) Overall dimension and mass.
(d) A detailed statement of the design of any capsule to be used.
(e) A statement of the tests that have been performed and their results, or evidence based on calculations, to show that the radioactive material is capable of meeting the performance standards, or other evidence that the

special form radioactive material or low dispersible radioactive material meets the applicable requirements of the Transport Regulations.

(f) A specification of the applicable management system, as required by para. 306 of the Transport Regulations.

(g) Any proposed pre-shipment actions for use in the consignment of special form radioactive material or low dispersible radioactive material.

INFORMATION TO BE INCLUDED IN APPLICATIONS FOR APPROVAL OF SHIPMENTS

I–5. The following information is to be included in applications for approval of shipments:

(a) General information:
  (i)   General description of the shipment from consignor to consignee, including loading, carriage and unloading of the consignment; stowage arrangements; storage in transit; and provisions for exclusive use, if applicable.
  (ii) Number of shipments.
  (iii) List of applicable national and international regulations and the edition of the Transport Regulations under which competent authority approval is sought.
  (iv) Specification of the applicable management system, radiation protection programme and emergency procedure.
  (v)   Radiation protection programme for shipments by special use vessels, in accordance with para. 576(a) of the Transport Regulations.
  (vi) Applicable package design certificates of approval (i.e. Type B(M) packages or packages containing fissile material).
(b) Administrative information, including the following:
  (i)   Name, address, telephone number and email address of the applicant;
  (ii)   Name, address, telephone number and email address of the consignor;
  (iii) Name, address, telephone number and email address of the consignee;
  (iv)   Name, address, telephone number and email address of the carrier;
  (v)   Type of shipment approval required, as specified in para. 825 of the Transport Regulations;
  (vi)   Identification mark of the competent authority, if previously allocated;
  (vii) Date of application;
  (viii) Date by which approval is desired.

(c)   Specific information for shipments described in para. 825(a)–(c) and (e) of the Transport Regulations, as required by para. 827 of the Transport Regulations, including the following:

(i)   The period of time, related to the shipment, for which the approval is sought.

(ii)   The actual radioactive contents, including the following:
— Radionuclides present;
— Total activity;
— Nature of emitted radiation;
— Heat output;
— Physical and chemical state;
— Quantity in mass units; for packages containing fissile material, the quantity of fissile material or fissile nuclides in mass units and the enrichment in percentage; for irradiated fuel, the burnup, irradiation time, cooling time and initial enrichment.

(iii)   The modes of transport.

(iv)   The type of conveyance and the probable or proposed route.

(v)   The details of how the precautions and administrative or operational controls, referred to in the certificates of approval for the package design, if applicable, issued under paras 810, 813 and 816 of the Transport Regulations, are to be put into effect.

(d)   Additional specific information for SCO-III shipments, as required by para. 827A of the Transport Regulations:

(i)   A statement of the respects in which, and the reasons why, the consignment is considered SCO-III (surface contaminated object of the group SCO-III).

(ii)   Justification for choosing SCO-III by demonstrating the following:
— No suitable packaging currently exists.
— Designing or constructing a packaging or segmenting the object is not practically, technically or economically feasible.
— No other viable alternative exists.

(iii)   A detailed description of the proposed radioactive contents with reference to their physical and chemical states and the nature of the radiation emitted.

(iv)   A detailed statement of the design of the SCO-III, including complete engineering drawings and schedules of materials and methods of manufacture.

(v)   All information necessary to satisfy the competent authority that the requirements of paras 520(e) and 522 of the Transport Regulations, if applicable, are satisfied.

(vi) A transport plan that describes various aspects of the shipment, as required by para. 520(e)(iii) of the Transport Regulations.

(vii) A specification of the applicable management system, as required in para. 306 of the Transport Regulations.

INFORMATION TO BE INCLUDED IN APPLICATIONS FOR APPROVAL OF SHIPMENTS UNDER SPECIAL ARRANGEMENT

I–6. The following information is to be included in applications for approval of design of shipments under special arrangement:

(a) General information:
  (i) General description of the shipment from consignor to consignee, including loading, carriage and unloading of the consignment; stowage arrangements; storage in transit; and provisions for exclusive use, if applicable.
  (ii) Number of shipments.
  (iii) List of applicable national and international regulations and the edition of the Transport Regulations under which competent authority approval is sought.
  (iv) Specification of the applicable management system, radiation protection programme and emergency procedure.
(b) Administrative information, including the following:
  (i) Name, address, telephone number and email address of the applicant;
  (ii) Name, address, telephone number and email address of the consignor;
  (iii) Name, address, telephone number and email address of the consignee;
  (iv) Name, address, telephone number and email address of the carrier;
  (v) Identification mark of the competent authority, if previously allocated;
  (vi) Date of application;
  (vii) Date by which approval is desired.
(c) Specific information as required by para. 830 of the Transport Regulations. Paragraph 830 of the Transport Regulations states:

> "An application for *approval* of *shipments* under *special arrangement* needs to include all the information necessary to satisfy the *competent authority* that the overall level of safety in transport is at least equivalent to that which would be provided if all the applicable requirements of

these Regulations had been met. The application shall also include:

> (a) A statement of the respects in which, and of the reasons why, the *shipment* cannot be made in full accordance with the applicable requirements;
> (b) A statement of any special precautions or special administrative or operational controls that are to be employed during transport to compensate for the failure to meet the applicable requirements."

This information also needs to include a detailed description of the radioactive material, the packaging and all the compensatory measures (i.e. technical, operational and administrative).

INFORMATION ON MANAGEMENT SYSTEMS

I–7. The extent of the management system will depend on the type of transport activities being performed by the organization. These activities include the design, manufacture, maintenance and repair of packaging, and the preparation, consignment, loading, carriage (including in-transit storage), shipment after storage, unloading and receipt at the final destination of loads of radioactive material and packages.

I–8. Although this subsection focuses on information on management systems to be included in applications for approval of design of packages, design of special form radioactive material and low dispersible radioactive material, and shipments and shipments under special arrangement, it provides comprehensive information that can be adapted to any of the transport activities mentioned above.

I–9. Information on management systems that is included with applications for approval might include the following information, in accordance with international, national or other standards acceptable to the competent authority:

(a) Management system organization:
  (i) Documentation of the formal structure of the organization by organization charts that identify each organizational element that functions under the management system.
  (ii) Documentation of the commitment of top management, stating that it is the policy of the organization to perform work on items important to transport safety in accordance with the management system.
(b) Management system programme:

(i)   Scope of the management system: a description of the measures established for identifying the following:
— The structures, systems and components covered by the management system;
— The approach to verifying that the applicable structures, systems and components meet the objectives of the management system.

(ii)   A description of the measures implemented to ensure the following:
— Activities important to safety are performed using specific instructions and specified equipment and under suitable environmental conditions.
— Management system manuals specify the designated responsibilities for implementation of activities important to safety.
— The user of the management system has established induction and training programmes to ensure that personnel performing activities important to safety are trained and qualified to perform those activities.

(c)   Package design control: A description of the measures implemented to ensure the following:

(i)   Cooperation among those responsible for preparing design documentation;

(ii)   Appropriate design analyses, including independent design verification;

(iii) Coordinating interfaces between involved personnel;

(iv)   Maintenance of lines of communication during the design process.

(d)   Procurement document control: A description of the measures implemented to control the preparation, review, concurrence and approval of all procurement documents.

(e)   Instructions, procedures and drawings: A description of the measures implemented to ensure the following:

(i)   Activities important to safety are prescribed and accomplished in accordance with current documented instructions, procedures or drawings, which have been approved by appropriate levels of management.

(ii)   All work activities are coordinated with management system personnel to ensure that the work controlling documents incorporate appropriate inspection and hold points to verify that initial work, planned work, effective repairs or rework have been performed satisfactorily.

(iii) Instructions, procedures and drawings include quantitative acceptance criteria (e.g. dimensions, tolerances, operating and regulatory limits) and qualitative acceptance criteria (e.g. workmanship samples) to verify that activities important to safety have been satisfactorily accomplished.

    (iv)   Written procedures address the use, management, storage and protection of electronic records and data.

    (v)   Information is maintained on the specific software applications and storage or computing hardware.

(f)   Document control:

    (i)   A description of the measures implemented to ensure that each of the documents under the control of the management system reflects its current status.

    (ii)   A description of controls established to ensure that all documents and changes thereto are adequately reviewed and approved prior to their issuance.

(g)   Control of purchased material, equipment and services: A description of the measures implemented to ensure that materials, equipment and services conform to procurement documents.

(h)   Identification and control of materials, parts and components: A description of the measures implemented to ensure that materials, parts and components, including partially fabricated assemblies, are adequately identified to preclude the use of incorrect or defective items.

(i)   Control of special processes[1]: A description of the measures implemented to ensure that special processes are controlled to ensure the following:

    (i)   Procedures, equipment and personnel are qualified in accordance with applicable codes, standards and specifications.

    (ii)   The operations are performed by qualified personnel and accomplished in accordance with written process or procedure sheets that direct the recording of evidence of verification.

    (iii) Qualification records of procedures, equipment and personnel are established, filed and kept current.

(j)   Internal inspection: A description of the measures established to ensure that the following activities concerning internal inspections are implemented:

    (i)   Inspection procedures, instructions or checklists are available for each work operation, where necessary to ensure quality.

    (ii)   Documents developed include methods for identifying characteristics and activities to be inspected, acceptance and rejection criteria, and the individuals or groups responsible for performing the inspection.

    (iii) Objective evidence of inspection results is recorded.

---

[1] Special processes may involve packaging maintenance by using certain processes (e.g. welding, heat treating) or non-destructive testing, or specific processes necessary to meet certificate of approval requirements.

(iv) Hold or witness points are identified.

(v) The appropriate personnel approve data to ensure that all inspection requirements have been satisfied.

(vi) The prerequisites to be satisfied prior to inspection are identified, including operator qualification and equipment calibration. Where sampling is used to verify acceptability of a group of items, the standard used as the basis for acceptance needs to be identified.

(vii) Inspectors are qualified in accordance with applicable codes, standards and training programmes.

(viii) Appropriate inspections are performed during various phases of operation, such as receiving inspections, in-process inspections, final inspections and maintenance inspections.

(k) Test control: A description of the measures established to ensure that applicable test programmes, including prototype qualification tests, production tests, proof tests and operational tests, are accomplished in accordance with written procedures. This includes measures established to ensure that modifications, repairs and replacements are tested in accordance with the original design and testing requirements.

(l) Control of measuring and test equipment: A description of the measures established to ensure that measurement and test equipment (e.g. gauges, fixtures, reference standards, devices used to measure product characteristics) is calibrated, adjusted and maintained at prescribed intervals or prior to use.

(m) Handling, storage and shipping control: A description of the measures established to ensure that cleaning, handling, storage and shipping are accomplished in accordance with design requirements to preclude damage or deterioration by environmental conditions such as temperature and humidity.

(n) Inspection, test and operating status: A description of the measures established to ensure that the status of inspections, tests and operating conditions (including maintenance of items) is known by organizations responsible for ensuring quality.

(o) Non-conforming materials, parts or components: A description of the measures established for controlling non-conforming items, including the following principal elements:

(i) Proper identification;

(ii) Segregation of discrepant or non-conforming items;

(iii) Disposition of the non-conforming items;

(iv) Evaluation of the non-conforming items.

(p) Corrective actions: A description of the measures established to ensure that the causes of conditions detrimental to quality (e.g. those resulting from failures, malfunctions, deficiencies, deviations, or defective material and equipment) are promptly identified and reported to appropriate levels of management. Also, a description of the measures established to obtain corrective actions from suppliers and ensure that follow-up actions are documented to verify that the corrective actions were implemented and effective.

(q) Management system records: A description of the measures established to ensure that management system records provide documentary evidence of the activities that affect quality and provide sufficient information to allow each record to be identified with the items or activities to which it applies. At a minimum, management system records might include the following information:

   (i)   Design, procurement, manufacturing and installation records;
   (ii)   Supplier evaluations;
   (iii) Non-conformance reports;
   (iv)   Results of inspections and tests;
   (v)   Failure analyses;
   (vi)   As-built drawings and specifications;
   (vii) Qualification of personnel, procedures and equipment;
   (viii) Calibration procedures;
   (ix)   Training and retraining records;
   (x)   Corrective action reports;
   (xi)   Records demonstrating evidence of operational capability;
   (xii) Records verifying repair, rework and replacement;
   (xiii) Audit plans, audit reports and corrective actions;
   (xiv) Records that are used as a baseline for maintenance.

(r) Audits: A description of the measures implemented to ensure that internal audits are performed that address the following elements:

   (i)   Assurance of authority and organizational independence of the auditors;
   (ii)   A commitment to adequate staffing levels, funding and facilities to implement the audit;
   (iii) Identification of audit personnel and their qualifications;
   (iv)   Provisions for reasonable and timely access of audit personnel to facilities and documents and to the qualified personnel necessary for performing audits;
   (v)   Use of established procedures and checklists;
   (vi)   Methods for reporting audit findings to the responsible management of both the audited and auditing organizations;
   (vii) Provisions for the audit team to gain access to levels of management that have responsibility and authority for corrective action;

(viii) Methods for verifying that effective corrective action has been accomplished on a timely basis.

69

REFERENCES TO ANNEX I

[I–1] INTERNATIONAL ATOMIC ENERGY AGENCY, Regulations for the Safe Transport of Radioactive Material, 2018 Edition, IAEA Safety Standards Series No. SSR-6 (Rev. 1), IAEA, Vienna (2018).
[I–2] INTERNATIONAL ATOMIC ENERGY AGENCY, Format and Content of the Package Design Safety Report for the Transport of Radioactive Material, IAEA Safety Standards Series No. SSG-66, IAEA, Vienna (2022).

Annex II

EXAMPLE TEMPLATES FOR CERTIFICATES OF APPROVAL

II–1. This annex provides example templates for certificates of approval by the competent authority for the following:

(a)   Design of packages;
(b)   Design of special form radioactive material and low dispersible radioactive material;
(c)   Shipments;
(d)   Shipments under special arrangement.

CERTIFICATE   OF   APPROVAL   FOR   DESIGN   OF PACKAGES   CONTAINING RADIOACTIVE MATERIAL

1. Expiry date of certificate.	2. Competent authority identification mark.

3. This certificate is issued on the basis of the application by:

[Name and address of the applicant]	[Reference to the application]

4. This is to certify that the design of the package described in the following meets the applicable requirements for [Type B(U), Type B(M), Type C package] [Type...package containing fissile material] in IAEA Safety Standards Series No. SSR-6 (Rev. 1), Regulations for the Safe Transport of Radioactive Material, 2018 Edition [hereinafter referred to as 'SSR-6 (Rev. 1)'], and in the regulations listed in Section 17 of this certificate.

This certificate does not relieve the consignor from compliance with any requirement of the government of any country through or into which the package will be transported.

Issue date.
[Signature of the certifying official(s)]
Address, telephone number and email address of the competent authority.

5. Package identification:

(a) Reproducible illustration not larger than 21 cm × 30 cm showing the make-up of the package.

(b) Packaging:
    (i)   Model name or number;
    (ii)   Description (e.g. use, dimensions, materials of manufacture, closures, penetrations, gross mass);
    (iii)   Reference to drawings or specification of design;
    (iv)   Description of the containment system.
(c) Radioactive contents (non-fissile):
    (i)   Radioisotopes;
    (ii)   Physical and chemical form (including special form radioactive material or low dispersible radioactive material, if applicable) and the mass in grams;
    (iii)   Maximum activity per package (including activities of the various isotopes);
    (iv)   Any restrictions on the radioactive contents that might not be obvious from the nature of the packaging.
(d) For designs of packages containing fissile material that require multilateral approval of the package design in accordance with para. 814 of SSR-6 (Rev. 1):
    (i)   Type and form of fissile material;
    (ii)   The maximum total mass of fissile nuclides or the mass for each fissile nuclide, when appropriate;
    (iii)   Description of the confinement system;
    (iv)   Criticality safety index;
    (v)   Reference to the documentation that demonstrates the criticality safety of the package;
    (vi)   Special features on the basis of which the absence of water from certain void spaces has been assumed in the criticality assessment;
    (vii)   Any allowance (based on para. 677(b) of SSR-6 (Rev. 1)) for a change in neutron multiplication assumed in the criticality assessment as a result of actual irradiation experience;
    (viii) The ambient temperature for which the package design has been approved.

6. References to certificates for alternative radioactive contents, other competent authority validation, or additional technical data or information.

7. Restrictions on the modes of transport.

8. If deemed appropriate, a statement authorizing shipment, where approval of shipment is required under para. 825 of SSR-6 (Rev. 1).

9. Specification of the management systems of the organizations involved in transport.

10. Operational controls for the preparation, loading, carriage, unloading and handling of the consignment, including any special stowage provisions for safe dissipation of heat.

11. Reference to information provided by the applicant relating to the use of the packaging or to specific actions to be taken prior to shipment.

12. A statement regarding ambient conditions assumed for the purposes of design, if these are not in accordance with those specified in paras 656, 657 and 666, as applicable, of SSR-6 (Rev. 1).

13. For Type B(M) packages, a statement specifying those prescriptions of paras 639, 655–657 and 660–666 of SSR-6 (Rev. 1) with which the package does not conform and any amplifying information that might be useful to other competent authorities.

14. For package designs subject to para. 820 of SSR-6 (Rev. 1), a statement specifying those requirements of the current regulations with which the package does not conform.

15. For packages containing more than 0.1 kg of uranium hexafluoride, a statement specifying those prescriptions of para. 634 of SSR-6 (Rev. 1) that apply, if any, and amplifying information that may be useful to other competent authorities.

16. Emergency arrangements deemed necessary by the competent authority.

17. Applicable regulations concerning the transport of radioactive material:

(a) Road:
(b) Rail:
(c) Sea:
(d) Inland waterways:
(e) Air:
(f) International:
(g) Other:

18. Table summarizing past and current revisions of the certificate of approval.

CERTIFICATE OF APPROVAL FOR DESIGN OF SPECIAL FORM RADIOACTIVE MATERIAL AND LOW DISPERSIBLE RADIOACTIVE MATERIAL

1. Expiry date of certificate.	2. Competent authority identification mark.

3. This certificate is issued on the basis of the application by:

[Name and address of the applicant]	[Reference to the application]

4. Identification of the special form radioactive material or low dispersible radioactive material (e.g. model name and number).

5. Description of the special form radioactive material or low dispersible radioactive material (means of encapsulation (if applicable), shape, dimensions).

6. Radioactive material (i.e. radionuclides, physical and chemical forms).	7. Maximum activity.
8. Design specifications (reference to drawings).	9. Specification of the applicable management system.

10. Specific actions to be taken prior to shipment.

11. This is to certify that the design of the special form radioactive material (or low dispersible radioactive material) described above meets the applicable requirements in IAEA Safety Standards Series No. SSR-6 (Rev. 1), Regulations for the Safe Transport of Radioactive Material, 2018 Edition, and in the regulations listed in Section 12 of this certificate.

Issue date.
[Signature of certifying official(s)]
Address, telephone number and email address of the competent authority.

12. Applicable regulations concerning the transport of radioactive material:

(a) Road:
(b) Rail:
(c) Sea:
(d) Inland waterways:
(e) Air:
(f) International:
(g) Other:

13. Table summarizing past and current revisions of the certificate of approval.

<table>
<tr><td colspan="2">CERTIFICATE OF APPROVAL FOR SHIPMENTS</td></tr>
<tr><td>1. Expiry date of certificate.</td><td>2. Competent authority identification mark.</td></tr>
<tr><td colspan="2">3. This certificate is issued on the basis of the application by:</td></tr>
<tr><td>[Name and address of the applicant]</td><td>[Reference to the application]</td></tr>
<tr><td colspan="2">4. This is to certify that the shipment of the radioactive material described in the following is designed to meet the applicable requirements for the shipment of the radioactive material in IAEA Safety Standards Series No. SSR-6 (Rev. 1), Regulations for the Safe Transport of Radioactive Material, 2018 Edition [hereinafter referred to as 'SSR-6 (Rev. 1)'] and in the regulations listed in Section 12 of this certificate.

This certificate does not relieve the consignor from compliance with any requirement of the government of any country through or into which the package will be transported.</td></tr>
<tr><td colspan="2">Issue date.
[Signature of the certifying official(s)]
Address, telephone number and email address of the competent authority.</td></tr>
<tr><td colspan="2">5. Identification of the applicable certificate(s) of approval of design.</td></tr>
<tr><td colspan="2">6. Specification of actual radioactive contents, including the following:

(a) Radioisotopes (including fissile material);
(b) Physical and chemical form (e.g. special form radioactive material, low dispersible radioactive material, or fissile material excepted under para. 417(f) of SSR-6 (Rev. 1), if applicable);
(c) Total activity per package and per conveyance (including activities of the various isotopes, if appropriate) and total mass in grams per package and conveyance;
(d) Total amount in grams of fissile material (or for each fissile nuclide, when appropriate) per package and per conveyance;
(e) Any restrictions on the radioactive contents that might not be obvious from the nature of the packaging.</td></tr>
<tr><td colspan="2">7. Restrictions on the modes of transport or types of conveyance and/or freight container, and any necessary routeing instructions.</td></tr>
<tr><td colspan="2">8. Specification of the management systems of the organizations involved in transport.</td></tr>
<tr><td colspan="2">9. Operational controls required for preparation, loading, carriage, unloading and handling of the consignment, including any special stowage provisions for the safe dissipation of heat or maintenance of criticality safety.</td></tr>
</table>

10. Reference to information provided by the applicant relating to specific actions to be taken prior to shipment.

---

11. Emergency arrangements deemed necessary by the competent authority.

---

12. Applicable regulations concerning the transport of radioactive material:

(a) Road:
(b) Rail:
(c) Sea:
(d) Inland waterways:
(e) Air:
(f) International:
(g) Other:

---

13. Table summarizing past and current revisions of the certificate of approval.

CERTIFICATE OF APPROVAL FOR SHIPMENTS UNDER SPECIAL ARRANGEMENT

1. Expiry date of certificate.	2. Competent authority identification mark.

3. This certificate is issued on the basis of the application by:

[Name and address of the applicant]	[Reference to the application]

4. This is to certify that the shipment of the radioactive material described in the following is designed to meet the applicable requirements for the shipment of the radioactive material under special arrangement in IAEA Safety Standards Series No. SSR-6 (Rev. 1), Regulations for the Safe Transport of Radioactive Material, 2018 Edition (hereinafter referred to as 'SSR-6 (Rev. 1)') and in the regulations listed in Section 16 of this certificate.

This certificate does not relieve the consignor from compliance with any requirement of the government of any country through or into which the package will be transported.

Issue date.
[Signature of the certifying official(s)]
Address, telephone number and email address of the competent authority.

5. Package identification:

(a) Reproducible illustration not larger than 21 cm × 30 cm showing the make-up of the package.
(b) Packaging:
    (i)   Model name or number;
    (ii)   Description (e.g. use, dimensions, materials of manufacture, closures, penetrations, gross mass);
    (iii)   Reference to drawings or a specification of the design;
    (iv)   Description of the containment system.
(c) Radioactive contents (non-fissile):
    (i)   Radioisotopes;
    (ii)   Physical and chemical form (including special form radioactive material, low dispersible radioactive material, or fissile material excepted under para. 417(f) of SSR-6 (Rev. 1), if applicable);
    (iii)   Maximum activity per package (including activities of the various isotopes, if appropriate) and total mass in grams per package;
    (iv)   Any restrictions on the radioactive contents that might not be obvious from the nature of the packaging.
(d) Additionally, for packages containing fissile material:
    (i)   Type and form of fissile material;
    (ii)   The maximum total mass of fissile nuclides or the mass for each fissile nuclide, when appropriate;
    (iii)   Description of the confinement system;
    (iv)   Criticality safety index;

(v)   Reference to the documentation that demonstrates the criticality safety of the package;

(vi)   Special features on the basis of which the absence of water from certain void spaces has been assumed in the criticality assessment;

(vii)   Any allowance (based on para. 677(b) of SSR-6 (Rev. 1)) for a change in neutron multiplication assumed in the criticality assessment as a result of actual irradiation experience;

(viii) The ambient temperature for which the special arrangement has been approved.

6. References to certificates for alternative radioactive contents, other competent authority validation, or additional technical data or information.

7. Mode(s) of transport and identification of carrier(s).

8. Restrictions on the modes of transport or types of conveyance and/or freight container, and any necessary routeing instructions.

9. Specification of the management systems of the organizations involved in transport.

10. Operational controls for the preparation, loading, carriage, unloading and handling of the consignment, including any special stowage provisions for the safe dissipation of heat.

11. Reference to information provided by the applicant relating to the use of the packaging or to specific actions to be taken prior to shipment.

12. Reasons for the special arrangement.

13. Compensatory measures to be applied as a result of the shipment being under special arrangement.

14. Ambient conditions assumed for purposes of design if these are not in accordance with those specified in paras 656, 657 and 666, as applicable, in SSR-6 (Rev. 1).

15. Emergency arrangements deemed necessary by the competent authority.

16. Applicable regulations concerning the transport of radioactive material:

(a) Road:
(b) Rail:
(c) Sea:
(d) Inland waterways:
(e) Air:
(f) International:
(g) Other:

17. Table summarizing past and current revisions of the certificate of approval.

Annex III

INSPECTIONS OF MANAGEMENT SYSTEMS AND
RELATED INSPECTION ACTIVITIES PERFORMED
BY THE COMPETENT AUTHORITY

III–1. The following is a list of general items which the competent authority may review and verify during inspections:

(a) The management of the organization has provided the necessary personnel and resources to implement an effective programme for compliance with IAEA Safety Standards Series No. SSR-6 (Rev. 1), Regulations for the Safe Transport of Radioactive Material, 2018 Edition [III–1] (hereinafter the 'Transport Regulations'). This programme needs to clearly identify the persons responsible for fulfilling the various specific requirements. There needs to be a clear delegation of authority by management to those responsible persons.

(b) The management has provided proper training to the persons responsible for implementing the programme for compliance with the Transport Regulations. Documentation of the training that has been provided needs to be submitted to the competent authority upon request.

(c) Established procedures are followed for the design and fabrication or the selection and procurement of packagings.

(d) The consignor is using the proper packaging for the specific contents of packages. The competent authority may perform direct examination of packages being prepared for shipment.

(e) The organization has in its possession all the required documentation in accordance with the Transport Regulations, including the relevant competent authority certificates and any associated instructions for handling, loading, storage, use and maintenance of the packaging (often given in the form of an instruction manual for the packaging).

(f) Established procedures are followed for the preparation and use of the package, in accordance with the certificate of approval, the instruction manual and related documents.

(g) Established procedures are followed for the proper marking and labelling of packages, in accordance with the Transport Regulations. This includes the proper determination and application of the correct transport index. When practicable, the competent authority may directly observe such actions.

(h)   Established procedures are followed, and appropriate and properly calibrated instruments are provided, to monitor packages for both radiation and contamination.

(i)   Established procedures are followed for the correct preparation and control of all relevant shipping documents and for providing (i) correct placarding of the carrier's vehicles, (ii) all the required documentation to carriers, and (iii) any required notification to the competent authorities of each country into which or through which the consignment is transported.

(j)   During transport, carriers perform any required actions relating to placarding, stowage and segregation of packages, particularly any administrative controls relating to exclusive use shipments, or supplementary operational controls as specified in the competent authority certificate.

(k)   The organization has established an appropriate radiation protection programme for its activities concerning the transport of radioactive material, and the programme is maintained, reviewed and complied with.

(l)   Procedures have been developed and implemented to respond to cases of non-compliance, appropriate investigative and corrective actions have been taken, and the necessary reporting and communicative action is being achieved.

(m)   The organization has developed and continues to maintain appropriate emergency arrangements and conducts exercises for emergency preparedness and response periodically, as appropriate.

III–2. Example procedures and checklists that can be used by competent authorities for their inspection activities are given in Annexes IV–IX. These procedures and checklists are not comprehensive and can be used as a starting point for a competent authority to develop its own procedures and checklists in accordance with the size and complexity of the industry and operations being inspected.

REFERENCE TO ANNEX III

[III–1]   INTERNATIONAL ATOMIC ENERGY AGENCY, Regulations for the Safe Transport of Radioactive Material, 2018 Edition, IAEA Safety Standards Series No. SSR-6 (Rev. 1), IAEA, Vienna (2018).

Annex IV

**EXAMPLE PROCEDURE FOR INSPECTING
A MANAGEMENT SYSTEM**

INSPECTION OF THE MANAGEMENT SYSTEM	Procedure No.:
	Revision No.:
	Page No.:
	Originator:
	Date:

CONTENTS

(1) Purpose
(2) Scope
(3) Definitions
(4) Responsibilities
(5) Procedure
(6) Records
(7) Declaration

Revision No.:	Approval date:	Authorized by:	Position:	Approved by:	Position:

1.   PURPOSE

1.1. To define the method used by the competent authority to perform inspections of the management system (in support of the compliance assurance programme developed by the competent authority in conformance with the requirements of IAEA Safety Standards Series No. SSR-6 (Rev. 1), Regulations for the Safe Transport of Radioactive Material, 2018 Edition [IV–1] (hereinafter the 'Transport Regulations')).

2.   SCOPE

2.1.   The procedure covers the inspection activities of the competent authority and its agents in connection with an inspection programme specified by a nominated person (denoted in this annex as the 'head of compliance') and agreed to by management. Inspection activities in addition to the planned programme of inspection can be arranged if requested by other organizational units of the competent authority. Inspection activities include the following:

(a)   Establishing whether the elements within the management system are properly documented;
(b)   Verifying through reviews and evaluation of documentary evidence that the management system is being adequately implemented;
(c)   Evaluating the adequacy, effectiveness and efficiency of the management system;
(d)   Identifying non-compliance, and requesting and verifying corrective actions.

3.   DEFINITIONS

3.1.   Inspection checklist: A list of the enquiries to be raised by the inspection team, which constitutes the inspection scope.

3.2.   Inspection matrix: A chart of the activities inspected and the standard(s) for management systems against which the activities have been inspected.

3.3.   Inspection plan: A timetable of inspecting activities.

3.4.   Inspector(s): The person(s) responsible for undertaking the inspection.

3.5.   Inspection: An inspection of the prescribed arrangements and their provisions against the requirements of international or national regulations.

3.6.   Corrective actions: Measures or actions taken to rectify non-compliance or to prevent any recurrence.

3.7. Non-compliance: An identified deviation or departure from the provisions of the specified standard for the management system or the arrangements prescribed under the management system.

3.8. Observation: A reportable deviation from good working practice that might give rise to a quality related problem.

3.9. Record of non-compliance or observation: (a) Recorded evidence, details and resulting corrective actions of non-compliance with the standards or procedures on which the inspection is based or (b) observation on the basis of which the management system has been found inadequate, which is noted during the inspection.

3.10. Inspection report: A document summarizing the inspection results, which includes the inspection findings and corrective actions to be undertaken, issued after the inspection by the competent authority to the organization being inspected.

3.11. Inspection of the management system: A systematic and independent examination to determine whether the activities in the management system and the related results comply with planned arrangements and whether these arrangements are implemented effectively and are suitable for achieving the objectives of the organization.

4.   RESPONSIBILITIES

4.1. The head of compliance is responsible for the management of all inspections of the management system and for inspection by the competent authority. The head of compliance is responsible for appointing the inspection team leader.

4.2. The team leader is responsible for the planning, preparation, documentation and reporting of all quality inspection activities. In undertaking inspections of the management system, the following are developed:

(a)   The inspection plan;
(b)   Checklists;
(c)   The inspection matrix;
(d)   The inspection report;
(e)   A statement of completion of the inspection.

4.3. The organization being inspected is responsible for implementing the corrective actions noted in documented requests for corrective action.

4.4. The team leader is responsible for verifying that requests for corrective action have been implemented.

5.   PROCEDURE

Inspection preparation

5.1.   The head of compliance, or the designate, prepares and issues to management an overall inspection programme. The programme is reviewed and updated periodically.

5.2.   The head of compliance selects the inspection team and nominates a team leader. The team leader may delegate preparatory and follow-up activities to team members. Team members other than observers have received training in appropriate inspection techniques.

5.3.   The team leader opens an inspection file (all commercial information being confidential), allocates a sequential reference number and arranges initial contact with the organization being inspected. If other government departments have an interest in the inspection, they may be informed in accordance with any extant interdepartmental agreements.

5.4.   The general arrangements and plans for the inspection are prepared by correspondence and, if necessary, by means of a pre-inspection meeting of the team leader and the organization being inspected.

5.5.   The team leader records the proposed inspection activities in an inspection plan. A questionnaire (inspection checklist) is used and covers the scope of the inspection to be undertaken.

5.6.   An inspection matrix is drawn up, reflecting the criteria of the codes or standards against which the organization being inspected will be inspected.

5.7.   The agreed dates for the inspection are confirmed by means of correspondence with the organization being inspected; other interested parties are also notified. In all instances, the notification includes the following points:

(a)   The date and time of the planned inspection;
(b)   Details of the inspection plan;
(c)   The name(s) of the inspector(s);
(d)   The agenda for the opening meeting.

5.8. Prior to the inspection, an inspectors meeting is convened, at which the inspection plan, the inspection checklist and the inspection matrix are discussed. Any other relevant information, such as the results of previous inspections or reviews, is included.

Performance of the inspection

5.9. The inspection is opened by a meeting between the inspection team and the representatives of the organization being inspected. The topics covered at the meeting include the following:

(a) Introduction;
(b) The purpose of the inspection;
(c) The inspection plan and the scope of the inspection;
(d) The interests of other government departments;
(e) The closing meeting.

5.10. The inspection is conducted objectively so as to establish whether the areas under examination have a satisfactory management system and whether the organization being inspected is adhering to it.

5.11. An inspection matrix is completed by each inspector, indicating the criteria that have been inspected. In the final inspection review, the team leader checks all the inspected criteria against the criteria of the respective codes and standards. Areas not inspected are then highlighted, and the team leader can decide what actions have to be taken. The completed inspection matrix is then included in the inspection record, which can be used when future inspections are planned, for example for criteria not inspected or areas considered weak.

5.12. Evidence and details of non-compliance regarding the standards or procedures on which the inspection is based are recorded. The record of non-compliance indicates whether the necessary corrective action needs to be taken immediately or within a given time. This record is signed by a representative of the organization being inspected to confirm that it is factual and correct. However, if the representative of the organization being inspected does not countersign the non-compliance record, it can still be considered to be admissible if the team leader so decides.

5.13. The team leader regularly reviews the progress of the inspection, discussing non-compliance, changes to the inspection plan (if necessary) and other topics. In a final review before the closing meeting, the non-compliance, observations and conclusions to be presented at the closing meeting are agreed on by the inspection team. Also, the inspection matrix is completed, recording the areas and topics covered during the inspection. (This completed inspection matrix is considered to constitute evidence of compliance with the Transport Regulations.)

Closing meeting

5.14. A closing meeting (as decided on at the opening meeting) is convened with the management of the organization being inspected and the inspection team. The team leader presents a balanced summary of the inspection undertaken, referring to the positive aspects emerging from the inspection, as well as to the points of non-compliance and the observations indicating where the management system has been found to be inadequate. Copies of the records on non-compliance and observations are presented to the organization being inspected.

5.15. The representatives of the organization being inspected are invited to comment on the findings; any disagreement or clarification concerning corrective actions is discussed and, if possible, resolved. The organization being inspected is advised by the team leader that a written report of the inspection will be sent by the competent authority in due course.

Inspection report

5.16. The inspectors prepare an inspection report that summarizes the inspection results and includes the findings of the inspection and the corrective actions to be undertaken. Further consultation with other government departments is held at this stage, if necessary. When considered appropriate by the head of compliance, an interim inspection report, covering only the findings, may be prepared and sent to the organization being inspected for prompt information.

5.17. The inspection report is sent to the organization being inspected, together with a cover letter referring to follow-up and verification of corrective actions. Organizations being inspected are requested to respond formally to the findings of the inspection, stating the timescale for the completion of corrective actions.

5.18. The progress of corrective actions is monitored by the team leader. If problems are encountered in connection with these actions, the head of compliance and management may also be involved in this process. Where necessary, follow-up reports are issued to inform the appropriate senior management of the organization being inspected that a potential problem remains. When the inspection is complete, the team leader confirms this in a letter to the organization being inspected. The team leader also checks that all necessary documentation and records are filed and indexed. Completion of the inspection is certified by a written statement of inspection completion, which is signed by the team leader.

6. RECORDS

6.1. The following inspection records are retained by the competent authority:

(a) Inspection programmes;
(b) Individual inspection files, containing inspection plans, inspection matrices, inspection reports, follow-up letters, correspondence and statements of completion of the inspection;
(c) Index of completed inspections.

7. DECLARATION

7.1. This procedure does not preclude the competent authority from any enforcement action deemed necessary in accordance with its management framework for enforcement.

REFERENCE TO ANNEX IV

[IV–1] INTERNATIONAL ATOMIC ENERGY AGENCY, Regulations for the Safe Transport of Radioactive Material, 2018 Edition, IAEA Safety Standards Series No. SSR-6 (Rev. 1), IAEA, Vienna (2018).

Annex V

EXAMPLE CHECKLIST FOR INSPECTING A MANAGEMENT SYSTEM

MANAGEMENT SYSTEM AND STRATEGIC PLANNING

1. Is there an established and appropriately documented management system?

2. Is the organization's policy and statement of authority with respect to the management system documented?

3. Does the management system fully identify those processes and activities covered by the management system and provide for their effective control?

4. Is involvement in and commitment to the management system and its objectives evident on the part of senior management?

5. Does the management system fully cover the activities undertaken by the organization? (These activities might include the design, manufacture, maintenance and repair of packagings, and the preparation, consigning, loading, carriage (including in-transit storage), shipment after storage, unloading and receipt at the final destination of loads of radioactive material and packages.)

6. Is there a defined organizational structure, and are the management responsibilities consistent with the size and complexity of the organization and its functions?

7. Are the functional responsibilities and levels of authority clearly defined at all levels within the organization?

8. How does the organization manage organizational changes to ensure that it remains effective and that quality and compliance remain unaffected?

9. Are the provisions of the management system commensurate with the complexity of the packaging or its components and with the degree of hazard associated with the material being transported (i.e. a graded approach)?

10. Is the management system subject to review and evaluation? If so, how frequently?

11. Who is responsible for reviewing the management system?

12. Does the management system review process provide for an appropriate scope and include all necessary inputs?

13. Does the review process include a confirmation that the management system demonstrates compliance with IAEA Safety Standards Series No. SSR-6 (Rev. 1), Regulations for the Safe Transport of Radioactive Material, 2018 Edition [V–1] (hereinafter the 'Transport Regulations') relating to the transport of radioactive material by the modes of transport used by the organization?

14. Does a radiation protection programme exist within the organization?

15. What processes are used to encourage and manage a safety culture within the organization?

16. Do the strategic plans of the organization include the development of its policies, objectives and processes?

DOCUMENTATION AND CONTROL OF RECORDS

17. Has the management system documentation been sufficiently well defined and are all essential documents supporting the effective and efficient operation of the management system in place?

18. Are there documented procedures to control all necessary documentation and records?

19. Do the procedures cover the preparation, approval and issuing of such documentation?

20. Has a system for the release, distribution and withdrawal of documents been established?

21. How are personnel made aware of changes to documents?

22. How are suppliers made aware of changes to documents?

23. Are applicable codes, standards and regulations updated as amendments are issued?

24. How are essential personnel made aware that amendments to documents, including codes and standards, have been issued?

25. Are copies of redundant or out of date documents suitably marked, withdrawn or destroyed?

26. Are superseded or redundant documents retained? If so, how are they controlled to prevent their accidental use?

27. Are changes to documents subject to review and approval:

(a) In accordance with documented procedures?
(b) By designated persons or organizations having relevant background information and knowledge and understanding of the original document?

28. How are incoming and/or external documents controlled?

29. Are records of changes to documentation retained?

30. Does the system cover the maintenance of essential records of the management system?

31. Does the system cover the identification, collection, indexing, filing, storage, maintenance and disposal of records?

32. Are records readily retrievable and maintained in a suitable environment?

33. Are retention periods for records defined?

34. Are records or logbooks available for each package and packaging?

35. Do logbooks contain the necessary information, such as movement or transport records, authorized modifications to the package, and operating and maintenance instructions?

36. Are records available for maintenance performed at other locations?

MANAGEMENT RESPONSIBILITY

37. Has a management representative been appointed and given appropriate authority to manage, monitor, evaluate and coordinate the management system?

38. Are measurable objectives of the management system established for relevant functions and at relevant levels within the organization?

SATISFACTION OF INTERESTED PARTIES

39. Are the interested parties clearly identified?

40. Are the needs and expectations of interested parties identified?

41. What processes exist within the organization to monitor and measure the satisfaction of interested parties?

RESOURCE MANAGEMENT

42. Is there a commitment to the timely identification and provision of necessary resources, including personnel, to meet the needs of the organization and regulatory requirements?

43. Are the following items described in processes and/or procedures?

    (a) Human resources;
    (b) Infrastructure and working environment;
    (c) Financial resources;
    (d) Involvement of individuals;
    (e) Management of information and knowledge.

TRAINING

44. How does the organization encourage the involvement and development of its personnel?

45. How is this measured?

46. Does the organization provide an appropriate training programme for all personnel involved in the transport of radioactive material?

47. How are training needs identified?

48. How are training records maintained?

INFORMATION AND KNOWLEDGE MANAGEMENT

49. How does the organization characterize, monitor and manage information and knowledge?

50. Is the procedure for information and knowledge management appropriate for the organization's transport related activities, and is it being followed?

COMMUNICATION AND INTERFACES

51. Does the organization provide for the effective communication of its policies, the needs of its customers and its regulatory needs to all personnel within the organization?

52. Are internal and external lines of communication established and defined in processes and procedures?

53. Are interfaces between relevant organizations established?

    (a) Are the interfaces between organizations, including the responsibilities of each organization, clearly defined in processes and procedures?
    (b) Are these interfaces regularly reviewed?

DEVELOPMENT OF PROCESSES

54. Is there evidence that the organization has developed the management and work processes associated with the transport activities of the organization (e.g. design and manufacture, maintenance and repair, assembly and disassembly, loading, handling, labelling, dispatch, carriage, receipt, unloading, and storage of packages and packagings, as appropriate)?

PROCESS MANAGEMENT AND CONTROL OF PRODUCT

55. Are common processes such as document control, non-conformance control and corrective actions, management review, and internal audits identified and adopted throughout the organization?

DESIGN CONTROL

56. Are sufficient measures in place to control the design process, and are they described in processes and procedures?

57. Have appropriate responsibilities been assigned for the whole design process?

58. Have suitable procedures been established for communicating design information, including changes, between the following?

    (a) Design disciplines;
    (b) Different units within the same organization;
    (c) External interfaces, including manufacturing and maintenance and repair facilities.

59. Is a graded approach to design used? If so, is each grade defined? For example:

    (a) Grade 1:
        (i) Are the relevant regulations, industrial standards and codes defined?
        (ii) Does the management system dictate that design verification is to be accomplished by:
            — Formal design review or prototype testing;
            — Calculations; or
            — Computer codes?
    (b) Grade 2:
        (i) Are the relevant regulations and industrial standards and codes defined?

(ii) Does the management system dictate that design verification is to be accomplished by:

— Calculations; or

— Computer codes?

(c) Grade 3: Does the design follow accepted engineering or industrial practice?

60. Have provisions been made to ensure that all necessary design input — including customers' needs — has been identified and included in the design process?

61. Is the design input documented in a way that permits adequate evaluation by technical personnel other than those persons performing the original design?

(a) Are such evaluations planned?

(b) Are such evaluations documented?

(c) Are such evaluations performed before submitting design information to the competent authority and to suppliers, or before commencing manufacture?

62. Have all acceptance and verification criteria been identified in design output and included in design input?

63. Is design output sufficiently defined to demonstrate its conformance with design input specifications?

64. What measures are established for the selection of, and for the review of the suitability of, materials, equipment and processes?

(a) Are such measures defined in procedures or instructions?

(b) Are such selections and reviews documented?

(c) Are such selections and reviews evaluated by technical personnel other than those performing the original design work?

65. Are there suitable arrangements for review of the design output to confirm the adequacy of the design?

(a) Are design reviews conducted?

(b) Are they planned and systematic?

(c) Are they documented?

(d) Do they include technical personnel other than those persons performing the design work?

(e) Are alternative calculational methods employed?

66. Are there appropriate arrangements for verification and subsequent validation of the adequacy of the design, such as a programme of model, prototype or full scale testing in accordance with the requirements of the Transport Regulations?

67. Do the design procedures provide for the control of changes, deviations and concessions regarding the design specifications?
(a) Are changes, deviations and concessions documented?
(b) Do such documents need authorization by the person responsible for the design?
(c) Do such documents, after authorization, state the justification for acceptance of such changes, deviations and concessions?
(d) Are suitable records retained?

68. Are design related changes to existing and in-service equipment covered by appropriate process controls?
(a) Is there a procedure for controlling in-service changes or modifications?
(b) Are such in-service changes or modifications documented?
(c) Are in-service changes or modifications subject to the approval of the person responsible for the design?
(d) Are the justifications for accepting in-service changes and modifications and the necessary actions documented?
(e) Is information concerning changes sent to:
(i) All affected persons and organizations?
(ii) All personnel or organizations holding the original design?
(f) Are suitable records retained?

MANAGEMENT SYSTEMS AND THE DIFFERENT PHASES OF TRANSPORT

69. Does the management system clearly identify the different phases of transport applicable to the organization — namely, design and manufacture, maintenance and repair, assembly and disassembly, loading, handling, labelling, dispatch, carriage, receipt, and unloading and storage of packages and/or packagings — and the interfaces between them?

PURCHASING

70. Have effective and efficient purchasing processes been described in procedures and suitable controls implemented?

71. Do the purchasing processes and procedures provide for all necessary purchase criteria to be identified and specified on purchase orders and documents?

72. Do the purchasing processes and procedures ensure that the relevant design documents and regulatory requirements are included or referenced in procurement documents?

73. Are purchasing arrangements commensurate with the importance or safety related aspects of the product being procured? Is a graded approach to products and suppliers being taken?

74.   Do the procurement documents specify that design and quality specifications be passed on to subsuppliers?

75.   Do the procurement documents specify that material traceability be maintained throughout fabrication and assembly, as necessary?

76.   Are suppliers selected and evaluated for their ability to supply products in accordance with specific criteria? Who conducts this evaluation, and is it recorded?

77.   How is the past performance of suppliers recorded?

78.   How often is a supplier assessed?

79.   Are audits of supplier management systems conducted as part of the evaluation process?

80.   Are supplier audits planned and documented?

81.   Do purchasing documents provide for adequate access by the purchaser and the competent authority to the plants of the suppliers and subsuppliers?

82.   What controls are established to ensure that purchased items conform to the specifications of procurement documents? Are such controls documented?

83.   Are suitable arrangements provided for the care and control of customer supplied materials or items?

IDENTIFICATION, TRACEABILITY AND PRESERVATION OF MATERIALS

84.   Are suitable arrangements in place to determine when identification and traceability of materials, items and software are necessary?

85.   Are the necessary identification and traceability of such items being achieved?

86.   Are measures established to control the handling, storage and shipping of materials both at initial delivery and during use in transport operations?

87.   Are suitable provisions in place for the preservation and protection of products, materials and packagings to ensure their fitness for use when needed?

PROCESS CONTROL

88.   Have all relevant processes, including those necessary for the management system (e.g. design, procurement, manufacturing, delivery processes, transport operations), been identified and their control provided for?

89.   Do process control arrangements, including procedures, instructions and drawings, include appropriate qualitative and quantitative acceptance criteria for determining that important activities have been satisfactorily accomplished?

90.   Are documents such as quality plans produced to support process control? Are such plans available?

91.   Are any design, production or other processes performed by subsuppliers controlled, and how is this done?

92.   How are process control arrangements and procedures reviewed, controlled and issued?

93.   Do subsuppliers use their own process control procedures (e.g. process procedures, work instructions) or those of the purchasing organization?

94.   Are inspections or other process control checks performed at defined points during the process (e.g. manufacturing, maintenance)?

95.   Are special processes (e.g. welding, non-destructive testing) controlled?

96.   How are these special processes controlled and monitored?

97.   Are only suitably qualified and experienced people used to control or perform special processes?

98.   Are all necessary controls or supporting processes available for controlling special processes (e.g. heat treatment)?

CONTROL OF INSPECTIONS, MEASUREMENTS AND TESTS

Inspection

99.   Have programmes for inspection of items and services been established?

100.   Do the procurement documents request that suppliers establish inspection programmes?

101. Have programmes for in-service inspections been established?

102. Who authorizes inspection programmes?

103. Are such inspections conducted by qualified personnel other than those persons performing the activities?

104. Have inspection procedures been established?

105. What processes ensure that non-conforming in-service items are removed from use until the situation is rectified?

106. Are inspection hold points defined?

107. How is it ensured that work does not proceed beyond a hold point?

Measurement and monitoring

108. What processes does the organization use to measure and monitor the characteristics of its packagings, packages and conveyances to verify that the needs of customers and regulatory requirements have been met?

109. What processes does the organization use to release a packaging, package or conveyance to its customer?

Testing

110. Have test programmes been established?

111. How is it ensured that the test programmes demonstrate the adequacy of the specification(s) and that all parts will perform satisfactorily in service?

112. Is testing conducted against written test procedures, and are the acceptance criteria specified?

113. Who evaluates the test results?

114. Does testing cover normal and accident conditions of transport?

115. Does the management system cover calibration and control of the measuring and test equipment?

116. Are calibration records available, and can they be traced back to a national standard?

117. Is measuring and test equipment calibrated, adjusted and maintained at prescribed intervals or prior to use?

118.   Is measuring and test equipment labelled or tagged to indicate calibration status?

119.   If equipment is found to be non-compliant, how is acceptance of the items reassessed?

120.   Are controls established for the handling, storage and use of equipment?

121.   How is inspection and test status identified, and is it maintained throughout the manufacture and use of an item?

PROCEDURES AND PROCESSES FOR CERTAIN ACTIVITIES RELATED TO PACKAGES

122.   Have systems and control processes for handling, labelling, dispatch, carriage, receipt, unloading and storage of packages and packagings been established?

123.   Do procedures and processes include controls for the following:

(a) The contents;
(b) Cleaning of packages;
(c) Preservation;
(d) Leaktightness;
(e) Dose rates and radioactive contamination;
(f) Turnaround and periodic inspection of package and packaging;
(g) Consumable and spare packaging components;
(h) Transport documents?

SELF-ASSESSMENT

124.   Has a programme been established to perform organizational self-assessments at all levels of management to evaluate the performance of work?

INDEPENDENT ASSESSMENT

125.   Is there a programme for internal and external audits?

126.   Is the audit programme documented?

127.   Are audits conducted by qualified persons who have not been involved in the activity being audited?

128.   Are effective corrective and preventive actions taken when the system is found to be incorrect?

129.   Are internal audit reports used for management system reviews?

130. Is the organization subjected to independent assessment by any of its interested parties?

131. How do the findings of any independent assessments compare with the findings made by internal auditing?

NON-CONFORMANCE AND CORRECTIVE AND PREVENTIVE ACTIONS

132. Is there an effective system for controlling non-conforming material?

133. Are the procedures for rework, use-as-is and repair of non-conforming material documented and acceptable?

134. Is the responsibility for the review and acceptance of non-conforming items specified?

135. Are accepted non-conformances reported to the purchaser and, if necessary, to the competent authority?

136. Does the system provide for the detection of inferior quality and for the correction of its causes?

137. Is adequate action taken to correct the causes of inferior quality (e.g. design faults, defective material)?

138. Are analyses made to identify trends towards material non-conformance?

139. Does corrective action extend to material supplied by subcontractors?

140. Are data analysis and material examination conducted on failed items to determine the extent and causes of defects?

141. Is there an effective system for registration of corrective and preventive actions and events?

142. Is the effectiveness of corrective and preventive actions reviewed and monitored?

IMPROVEMENT

143. Are appropriate arrangements in place to review and confirm that the needs of customers and other interested parties are being met?

144. Does the organization have a process or procedure for continuous improvement, and is it being adequately implemented?

REFERENCE TO ANNEX V

[V–1]   INTERNATIONAL ATOMIC ENERGY AGENCY, Regulations for the Safe Transport of Radioactive Material, 2018 Edition, IAEA Safety Standards Series No. SSR-6 (Rev. 1), IAEA, Vienna (2018).

Annex VI

EXAMPLE CHECKLIST FOR INSPECTING CONSIGNORS

Inspection details				
Names of inspectors:				
Inspection reference files:				
Date/time:		Location:		
Company details and organization				
Name of company:		Address:		
Telephone:		Fax:		
Email:		Web site:		
Names of persons met				
Name	Title/function	Telephone		Email
List of packages				
Model	Manufacturer of packaging	Type	Reference of certificate of approval or documentation of compliance	Serial numbers

Activities performed by the consignor	
Activity	Comments
Receipt of radioactive material	
Package or special form radioactive material design	
Package manufacturing	
Radioactive material classification	
Selection of package type and of package design	
Preparation and/or handling of packages, loading, unloading, stowage of packages on conveyance	
Transport	
Maintenance or repair of packaging	
Does the consignor subcontract any of the above activities associated with the transport of radioactive material? (Identify what activities are subcontracted)	

Subject/inspection aspect	Paragraphs in IAEA Safety Standards Series No. SSR-6 (Rev. 1), 2018 Edition	Compliance			Comments
		Yes	No	n.a.[a]	
Company details and organization					
Does the consignor adequately define the interfaces with subcontractors and the respective responsibilities? (Identify how.)	306				
Does the consignor perform a previous evaluation of subcontractors as service suppliers? (Identify the applicable procedure.)	306				
Does the consignor have a procedure to cover the relationship with suppliers? (The relationship could be written in a specific accordance document, not necessarily in procedures.)	306				
Does the consignor have a list of approved suppliers? (Ask for and check some suppliers' documentation. Verify that the evaluation is in accordance with the procedure.)	306				
Does the consignor undertake periodic inspections of the subcontractor's activities? (Check procedures and records on these inspections.)	306				

Do the suppliers comply with other requirements like specific licences (e.g. carrier's registration or authorization, laboratory's authorizations)?	306 National regulations				
Are there written procedures to cover transport activities? (If, yes, identify them.)					
Is the content of those procedures in compliance with the applicable Transport Regulations (see next item) as well as with the certificate of approval and the safety analysis report of the package? (Whenever possible, check the fulfilment of the procedures.)	545–561 306				
Awareness of applicable Transport Regulations					
Is the company aware of the latest edition of the applicable modal, international and national regulations?	306				
Does the company hold a copy or copies of those regulations? (List those held.)	306				
How are copies controlled and updated? (Is a document system applied?)	306				

Subject/inspection aspect	Paragraphs in IAEA Safety Standards Series No. SSR-6 (Rev. 1), 2018 Edition	Compliance			Comments
		Yes	No	n.a.[a]	
Are the transport documents retained for a minimum of three months?	555				
Types of transport and package					
Modes of transport generally used by the consignor: (Identify the more usual consignments, consignees and transport routes.)					
By road					
By rail					
By air					
By sea					
By inland waterways					
Types of package used by the consignor:					
Excepted packages					

Industrial packages					
Type A					
Type B					
Type C					
Unpackaged radioactive material					
Other dangerous properties of contents:					
Toxicity (uranium hexafluoride)					
Fissile					
Others					
Is the radioactive material transported under exclusive use?					

Subject/inspection aspect	Paragraphs in IAEA Safety Standards Series No. SSR-6 (Rev. 1), 2018 Edition	Compliance			Comments
		Yes	No	n.a.[a]	
Evidence of conformity of the packages					
Radioactive material classification:					
Does the consignor do the classification?	401, 546, 408–434				
If yes, does the consignor have procedures for this activity? (Note references.)	306				
If not, does the consignor conduct any control over the classification process? (Identify the procedure and the way the consignor ensures this control: verification, inspection, calculation validation.)	306				
For special form radioactive material or low dispersible radioactive material, are the certificates of approval available? Are they still valid?	561, 556				

Is the radioactive material transported as 'fissile excepted'? (Identify the criteria used and the procedures applied by the consignor to confirm the criteria fulfilment.)	Table 1, footnote 'b', 417, 546(j)				
Packages:					
Does the consignor have procedures for selecting the packaging suitable for the radioactive material to be transported? (Identify the package designs, the number of packagings of each design used and their suppliers.)	306 401, 408–434				
For package designs subject to approval, are the certificates of approval in force in the possession of the consignor? Are they still valid?	561, 556 802				
Has the consignor implemented a procedure to be informed about changes of certificates of approval?	306				
For packages whose design is not subject to approval, does the consignor have in their possession the documentary evidence of the compliance of the package design? (Identify the documentation presented.) Is this documentation still valid (i.e. no design change)?	801				
Is the general condition of the packagings adequate?	502, 503				

Subject/inspection aspect	Paragraphs in IAEA Safety Standards Series No. SSR-6 (Rev. 1), 2018 Edition	Compliance			Comments
		Yes	No	n.a.[a]	
Are the components of the packagings in good condition?	502, 503				
Are the packagings and their components in accordance with the package design?	502, 503				
Is the method of marking packages adequate?	507, 531–537, 545, 547				
Is the method of labelling packages adequate?	507, 538–542, 545, 547				
Are the radiological measures conducted on the packages in accordance with regulations?	508, 509, 516, 523–524A, 526–529				
Package maintenance or repair (use Annex IX)					
Operating and handling processes					

Question	Reference				
Does the consignor possess the operation procedure referenced in the certificate of approval or in the compliance documentation? (Verify how the requirements included in those documents are transferred to consignor's instructions or procedures.)	561				
Does the consignor fulfil the predefined inspection requirements before each shipment?	502, 503				
Does the consignor fulfil inspection requirements before the first use of a packaging?	501				
Package marking and labelling					
Do the procedures include requirements for marking and labelling activities? (Check.)	531–542				
Is the marking on the packages in accordance with the Transport Regulations?	531–537				
Is the methodology for the determination of the transport index clearly defined and in accordance with the Transport Regulations?	523, 524				
Whenever possible, it is useful to do visual inspections.					

Subject/inspection aspect	Paragraphs in IAEA Safety Standards Series No. SSR-6 (Rev. 1), 2018 Edition	Compliance			Comments
		Yes	No	n.a.[a]	
Transport documentation and notification					
Do the procedures include the documentation required, and are they in compliance with the Transport Regulations? (Check for different modes; check shipment records and verify that they are in accordance with the procedures.)	546–556				
Does the documentation include: — The list of information provided in para. 546(a)–(n) of the Transport Regulations? — The consignor's declaration? — Name and address of the consignor? — Name and address of the consignee?	546 (a)–(n) 547–553 546 546				
Does the consignor provide supplementary transport requirements (e.g. handling, stowage, temperature measurements, if necessary)?	554(a)				
Does the consignor provide restrictions on the mode of transport or conveyance, and any necessary routing instructions? (If necessary.)	554(b)				

Question	Reference				
Does the consignor provide instructions on the mixed loading prohibition?	506, 507				
Does the consignor provide emergency arrangements appropriate to the consignment?	554(c)				
Is the documentation language used in accordance with the Transport Regulations?	554				
Does the consignor fulfil the notification requirements?	557, 558				
Radiation protection					
Does the consignor have a radiation protection programme (RPP)? If yes, record reference.	302				
Is the RPP maintained up to date?	301				
Is there adequate documentary evidence of the RPP?	302				
Is there a brief description of the operations?	302				
Are the responsibilities for radiation protection within the organization well defined?	302				

Subject/inspection aspect	Paragraphs in IAEA Safety Standards Series No. SSR-6 (Rev. 1), 2018 Edition	Compliance			Comments
		Yes	No	n.a.[a]	
Is a person assigned by the company to have overall responsibility for the RPP? (Identify who, which department, and that person's responsibility.)	302				
Is that person responsible for the following areas? — Training; — Implementation of work procedures; — Assessment of workers' exposures. (If not, identify the responsible individual.)	311 306 301, 303				
Are the working instructions and procedures in place adequate to optimize doses? (Identify the procedures implemented.)	301, 302				
Is there a structured and systematic approach to dose assessment?	301, 303				
Have dose assessments been undertaken? (Identify the procedure applied for the assessments.)	301–303				
Are records of individual monitoring of workers maintained, if required?	303				

Has radiological surveillance been undertaken? (If yes, describe; if no, justify.)	303				
Are the results of radiological surveillance recorded? (Check records.)	303				
Are contamination checks performed? (Describe method; check records.)	301, 508 509, 512				
Are contamination check records kept?	306				
Does the company know the applicable limits for dose rates and contamination?	508–514 526–529				
Is there a protocol for cases of non-compliance with the above limits for dose rates and contamination?	309				
Does the programme consider appropriate segregation distances between packages and areas regularly occupied by members of the public and/or workers?	562, 563, 506				
Are storage areas shielded in accordance with the RPP?	301, 562				
Is shielding used on the vehicles and freight containers?	566				
Is segregation used on the conveyances?	566–569				

Subject/inspection aspect	Paragraphs in IAEA Safety Standards Series No. SSR-6 (Rev. 1), 2018 Edition	Compliance			Comments
		Yes	No	n.a.[a]	
Are the necessary radiation protection measures and the optimization principle applied for the preparation of packages and their transport from in-transit storage to the loading area?	301				
Does the RPP incorporate requirements relating to emergency response in the event of a nuclear or radiological emergency during the transport of radioactive material?	302, 304, 305				
Does the RPP include training? (Identify training programmes, contents, initial and periodic training frequencies, and who performs the training.)	302, 311				
Is the training documented? (Identify how and check records.)	311, 314				
Are radiation monitoring devices: — Available? — Appropriate for the measurements to be taken? — Calibrated?	306				

Emergency arrangements					
Is a person assigned by the company to have overall responsibility for emergency preparedness and response? (Identify who and in which department, and the person's responsibility.)	304–306				
Are sufficient resources provided in case of emergency (e.g. on the site or during transport, relations with service suppliers as carriers)? (Identify.)	304–306				
Does the consignor have emergency arrangements appropriate to the consignment in place, and do those arrangements include details about contact points in case of emergency? (Identify, possibly in the RPP.)	304–306, 554(c)				
Do these provisions consider potential events that might happen during transport activities? (Identify, possibly in the RPP.)	304, 305				
Are these provisions regularly reviewed? (Check how the emergency provisions are implemented and maintained, if operative experience is considered.)	305, 306				

Subject/inspection aspect	Paragraphs in IAEA Safety Standards Series No. SSR-6 (Rev. 1), 2018 Edition	Compliance			Comments
		Yes	No	n.a.[a]	
Has there been a recent emergency? (Whenever possible, check the fulfilment of the emergency provisions during the emergency.)					
Training					
Does the consignor provide an appropriate training programme for all personnel involved in the transport of radioactive material?	311–315				
Does the consignor maintain records of the training and competence?	314				
Management system (use Annexes IV and V)					

[a] n.a.: not applicable.

Annex VII

EXAMPLE CHECKLIST FOR INSPECTING CARRIERS

Inspection details			
Names of inspectors:			
Inspection reference files:			
Date/time:		Location:	
Company details and organization			
Name of company:		Address:	
Telephone:		Fax:	
Email:		Web site:	
Names of persons met			
Name	Title/function	Telephone	Email
List of packages			

Model	Manufacturer of packaging	Type	Reference of certificate of approval or documentation of compliance	Serial numbers

Activities performed by the carrier	
Activity	Comments
Number of personnel involved with radioactive material transport and their status	
Percentage of business involving radioactive material transport	
Frequency of radioactive material transport (per month)	
Is there radioactive material transport in-house?	
Which modes of transport does the company use? (Road, rail, inland waterway, sea, air)	

Subject/inspection aspect	Paragraphs in IAEA Safety Standards Series No. SSR-6 (Rev. 1), 2018 Edition	Compliance			Comments
		Yes	No	n.a.[a]	
Awareness of applicable Transport Regulations					
Is the company aware of the latest edition of the applicable modal, international and national regulations?	306				
Does the company hold a copy or copies? (List those held.)	306				
How are copies controlled and updated? (Is a document system applied?)	306				
Radiation protection					
Does the carrier have a radiation protection programme (RPP)? If yes, record reference.	302				
Is the RPP maintained up to date?	301				
Is there adequate documentary evidence of the RPP?	302				
Is there a brief description of the operations?	302				

Subject/inspection aspect	Paragraphs in IAEA Safety Standards Series No. SSR-6 (Rev. 1), 2018 Edition	Compliance			Comments
		Yes	No	n.a.[a]	
Are the responsibilities for radiation protection within the organization well defined?	302				
Is a person assigned by the company to have overall responsibility for the RPP? (Identify who, which department, and that person's responsibility.)	302				
Are the working instructions and procedures in place adequate to optimize doses? (Identify the procedures implemented.)	301, 302				
Is there a structured and systematic approach to dose assessment?	301, 303				
Have dose assessments been undertaken? (Identify the procedure applied for the assessments.)	301–303				
Are records of individual monitoring of workers maintained, if required?	303				
Has radiological surveillance been undertaken? (If yes, describe; if no, justify.)	303				

Are the results of radiological surveillance recorded? (Check records.)	303				
Are contamination checks performed? (Describe method; check records.)	301, 505, 508, 509, 512, 513				
Are contamination check records kept?	306				
Does the company know the applicable limits for dose rates and contamination?	508–514 526–528, 566				
Is there a protocol for cases of non-compliance with the above limits for dose rates and contamination?	309				
Does the carrier maintain appropriate segregation distances between packages and areas regularly occupied by members of the public and/or workers in accordance with the RPP?	562, 563, 506				
Are storage areas shielded in accordance with the RPP?	301, 562				
Is shielding used on the vehicles and freight containers?	566				
Is segregation used on the conveyances?	566–569				
Are the necessary radiation protection measures and the optimization principle applied for the loading and unloading of conveyances?	301–302				

Subject/inspection aspect	Paragraphs in IAEA Safety Standards Series No. SSR-6 (Rev. 1), 2018 Edition	Compliance			Comments
		Yes	No	n.a.[a]	
Does the RPP include training? (Identify training programmes, contents, initial and periodic training frequencies, and who performs the training.)	302, 311				
Is training documented? (Identify how and check records.)	311, 314				
Are radiation monitoring devices: — Available? — Appropriate for the measurements to be taken? — Calibrated?	306				
Emergency arrangements					
Is a person assigned by the company to have overall responsibility for emergency preparedness and response? (Identify who and in which department and the person's responsibility.)	304–306				

Are sufficient resources provided in case of emergency (e.g. on the site or during transport, relations with consigners)? (Identify.)	304–306				
Does the carrier have emergency arrangements appropriate to the consignment in place, and do those arrangements include details about contact points in case of emergency?	304, 305				
Are the emergency procedures tested? (Provide details.)	306				
Driver/company					
Is the vehicle crew supplied with written instructions regarding emergency procedures?	554				
Have arrangements and procedures been established regarding packages that are damaged or leaking, and do those arrangements and procedures include the identification and assessment of contamination and dose rates associated with, and protective measures related to, such packages?	510, 511				
Does the crew have instructions or procedures to cover any transshipment, segregation or en route storage requirements?	554, 562, 563				

Subject/inspection aspect	Paragraphs in IAEA Safety Standards Series No. SSR-6 (Rev. 1), 2018 Edition	Compliance			Comments
		Yes	No	n.a.[a]	
Are accumulations of packages on conveyances monitored for dose rates, transport index and criticality safety index?	566, table 10, 569, table 11				
Is the mixed loading prohibition verified?	506, 507				
Does the company know the applicable conveyance activity limits for low specific activity (LSA) and surface contaminated object (SCO) packages?	522, table 6				
If a consignment has been transported under special arrangement, has the carrier implemented all relevant compensatory measures related to its carriage? Have arrangements been made concerning relevant compensatory measures for planned shipments under special arrangement?	310				
Training					
Does the company provide an appropriate training programme for all personnel involved in the transport of radioactive material?	311–315				

Does the company maintain records of the training and competence?	314				
Does the driver have any necessary documents to confirm their proficiency in handling radioactive material in accordance with national regulations (e.g. training certificate, driving licence)?	National regulations				
Transport documentation					
Is the driver supplied with all required transport documentation?	554, 584, 585				
Are the transport documents retained for a minimum of three months?	587				
Package and material transport activity					
Package and material type: The company will have, use or carry one or more of the following. (Enter the number for each type of package carried each month.)					

Excepted	IP 1, 2 or 3	Type A	Type B (state which subtype)	Special form material	Special arrangement	Fissile

Subject/inspection aspect	Paragraphs in IAEA Safety Standards Series No. SSR-6 (Rev. 1), 2018 Edition	Compliance			Comments
		Yes	No	n.a.[a]	
For Type B and Type C packages, and special form, low dispersible and fissile materials					
Does the company ask the consignor to make available copies of the package and/or material certificates of approval?					
Shipment approval certificates					
Is a procedure in place to meet shipment approval requirements, if necessary?	825, 829				
Are there shipment certificates of approval in place? (If yes, enter the certificate/authorization number.)	825, 829				
Placarding, fire extinguishers, miscellaneous equipment and stowage					
For road and rail: Are the vehicles correctly placarded?	571, 572				
Are large freight containers carrying unpackaged LSA-I material or SCO-I or packages other than excepted packages and tanks placarded?	543, 544				

Are vehicles subject to a maintenance programme?	306				
Are records kept of vehicle maintenance?	306				
Are all tie-down and anchorage systems of the vehicle subject to regular testing?	306				
Have arrangements and procedures been established that take account of the requirements relating to the surface heat flux of the package or overpack during carriage and stowage?	565				
For road: Do the fire extinguishers carried comply with the existing national provisions?	National regulations				
For road: Does other miscellaneous equipment carried comply with the existing national provisions?	National regulations				
Management system (use Annexes IV and V)					

[a] n.a.: not applicable.

EXAMPLE CHECKLIST FOR INSPECTING THE MANUFACTURING OF PACKAGINGS

Inspection details			
Names of inspectors:			
Inspection reference files:			
Date/time:		Location:	
Company details and organization			
Name of company:		Address:	
Telephone:		Fax:	
Email:		Web site:	
Names of persons met			
Name	Title/function	Telephone	Email

List of packages				
Model	Manufacturer of packaging	Type	Reference of certificate of approval or documentation of compliance	Serial numbers

Subject/inspection aspect	Paragraphs in IAEA Safety Standards Series No. SSR-6 (Rev. 1), 2018 Edition	Compliance			Comments
		Yes	No	n.a.[a]	
Management system (use Annexes IV and V)					
Management of resources					
Are human resources in development, manufacturing and quality assurance periodically evaluated for the work to be done by the company?	306				
Does the company provide an adequate training programme for its personnel? Is the staff sufficiently trained (e.g. qualification and competence preservation, knowledge of rules and standards, guidelines and state of the art)? (Ask for documentation of staff qualification.)	311–315				
Are tools and machines properly controlled, maintained and calibrated?	306				
Production and manufacturing of packagings					
Are the responsibilities for different production steps clearly stated?	306				

Subject/inspection aspect	Paragraphs in IAEA Safety Standards Series No. SSR-6 (Rev. 1), 2018 Edition	Compliance			Comments
		Yes	No	n.a.[a]	
Are specifications (e.g drawings, material used for constructing packagings) up to date and available to relevant personnel?	306				
Do all drawings used conform to those specified in the relevant certificate of approval of the competent authority or other compliance document?	306, 838				
Is there a valid and internally approved fabrication and test sequence plan?	306				
Are the realized test steps documented in the fabrication and test sequence plan?	306				
Has the manufacture been undertaken in accordance with the approved design specifications?	501				
Are the components of the packaging classified accordingly?	306				

Is the production of classified components documented accordingly? (Describe how the production of classified components is witnessed and documented.)	306				
Is the qualification of subcontractors monitored during procurement? Are there supporting documents?	306				
Only for competent authority approved packages: Are the fabrication and test plan organized with hold points and quality checks, and are they sufficiently documented? Only for non-competent authority approved packages: Is the organization of accompanying checks during manufacturing sufficient?	306, 801				
Are there compliance checks regarding specifications of the materials needed for production? (Ask for a list of material suppliers.)	306				
Are there certificates for materials in accordance with classified packaging components?	306				
Are the used materials traceable?	306				
Are the materials adequately stored and tested to ensure conformance with specifications?	306				
Are measuring and monitoring devices controlled?	306				

Subject/inspection aspect	Paragraphs in IAEA Safety Standards Series No. SSR-6 (Rev. 1), 2018 Edition	Compliance			Comments
		Yes	No	n.a.[a]	
Is the measuring and test equipment calibrated?	306				
Are measures established to handle deviations and/or changes?	306				
Only for competent authority approved packages: Does the manufacturer have a procedure to inform the competent authority about deviations and changes that have an impact on safety?	306				
Inspection before commissioning					
Do all manufactured packagings undergo the required acceptance inspections to ensure compliance with the design specifications?	501				
Is the package marked permanently?	531–536A				
Is the date of the next periodic inspection clearly visible?					
Are the results of inspections documented?	306				

Is there a control of completeness of documentation?	306				
Operation and maintenance of packagings (if applicable)					
Are the documents for operation of packages (instructions for use and maintenance) forwarded to the operator? (Describe how.)					
Is it ensured that the operator obtains instructions for use and maintenance of the packaging?	501–503				
Management of changes and improvement					
Are there procedures for ensuring feedback on operational experience of delivered packagings?	306				
Are changes in regulations and standards tracked? Are existing documents updated accordingly?	306				
Are changes in the design tracked? Are existing documents updated accordingly?	306				
Are deviation reports systematically evaluated and appropriate corrective and preventive measures implemented?	306				

[a] n.a.: not applicable.

Annex IX

EXAMPLE CHECKLIST FOR INSPECTING MAINTENANCE OPERATIONS

Inspection details				
Names of inspectors:				
Inspection reference files:				
Date/time:		Location:		
Company details and organization				
Name of company:		Address:		
Telephone:		Fax:		
Email:		Web site:		
Names of persons met				
Name		Title/function	Telephone	Email
List of packages				
Model	Manufacturer of packaging	Type	Reference of certificate of approval or documentation of compliance	Serial numbers

Subject/inspection aspect	Paragraphs in IAEA Safety Standards Series No. SSR-6 (Rev. 1), 2018 Edition	Compliance			Comments
		Yes	No	n.a.[a]	
Management system (use Annexes IV and V)					
Instructions for maintenance operations					
Are there instructions, procedures, plans or drawings for maintenance operations for each type of package?	306				
Are there procedures relating to periodic maintenance referenced in the certificate of approval or in the compliance documentation for each type of package?	306, 801, 838				
Are specified maintenance operations performed in due time and in accordance with the certificate of approval or compliance documentation for each type of package?	306, 801, 838				
Are records kept of maintenance operations?	306				
Are these records or logbooks correctly completed, verified or certified by authorized personnel?	306				

Subject/inspection aspect	Paragraphs in IAEA Safety Standards Series No. SSR-6 (Rev. 1), 2018 Edition	Compliance			Comments
		Yes	No	n.a.[a]	
Regulations					
Are the organization and personnel involved in the transport of radioactive material aware of the regulatory requirements?	312				
Resources					
Are the defined roles and responsibilities adequately resourced?	306				
Do the tools and equipment (in good condition and calibrated) comply with the relevant regulations?	306				
Training					
Does the company provide an adequate training programme for its personnel?	313				
Does the company maintain records of the training and qualifications of its personnel?	314				

Documentation, control of documents and records					
Is all requisite documentation completed and recorded by designated personnel?	306				
Are the necessary documents kept as records?	306				
Maintenance operations: controls, tests and inspections					
Does the company or facility have the necessary permits or licences for use or maintenance operations of packages and packaging?	National regulations				
Have the maintenance operations been undertaken in accordance with the package or packaging specifications?	306, 801, 838				
Is evidence available to show that specified controls, tests and inspections have been performed?	306				
Are there specific procedures or instructions to evaluate if repairs may affect the requirements defined for the package design in the certificate of approval and/or the package design safety report?	306, 838				

Subject/inspection aspect	Paragraphs in IAEA Safety Standards Series No. SSR-6 (Rev. 1), 2018 Edition	Compliance			Comments
		Yes	No	n.a.[a]	
Radiation protection					
Is there an adequate radiation protection programme (e.g. dose evaluation, optimization, radiological surveillance, radiation protection procedures)?	302				
Is the radiation protection programme periodically reviewed?	302, 306				

[a] n.a.: not applicable.

138

CONTRIBUTORS TO DRAFTING AND REVIEW

Badr, M.A.H.A.A.	Egyptian Nuclear and Radiological Regulatory Authority, Egypt
Buchelnikov, A.	State Atomic Energy Corporation, Russian Federation
Ershov, V.	State Atomic Energy Corporation, Russian Federation
Féron, F.	Nuclear Safety Authority, France
Nitsche, F.	Consultant, Germany
Reber, E.	International Atomic Energy Agency
Sahyun, A.	Nuclear and Energy Research Institute, Brazil
Tapp, J.	Nuclear Regulatory Commission, United States of America
Vaclav, J.	Nuclear Regulatory Authority of the Slovak Republic, Slovakia

ORDERING LOCALLY

IAEA priced publications may be purchased from the sources listed below or from major local booksellers.

Orders for unpriced publications should be made directly to the IAEA. The contact details are given at the end of this list.

NORTH AMERICA

Bernan / Rowman & Littlefield
15250 NBN Way, Blue Ridge Summit, PA 17214, USA
Telephone: +1 800 462 6420 • Fax: +1 800 338 4550
Email: orders@rowman.com • Web site: www.rowman.com/bernan

REST OF WORLD

Please contact your preferred local supplier, or our lead distributor:

Eurospan Group
Gray's Inn House
127 Clerkenwell Road
London EC1R 5DB
United Kingdom

Trade orders and enquiries:
Telephone: +44 (0)176 760 4972 • Fax: +44 (0)176 760 1640
Email: eurospan@turpin-distribution.com

Individual orders:
www.eurospanbookstore.com/iaea

For further information:
Telephone: +44 (0)207 240 0856 • Fax: +44 (0)207 379 0609
Email: info@eurospangroup.com • Web site: www.eurospangroup.com

Orders for both priced and unpriced publications may be addressed directly to:
Marketing and Sales Unit
International Atomic Energy Agency
Vienna International Centre, PO Box 100, 1400 Vienna, Austria
Telephone: +43 1 2600 22529 or 22530 • Fax: +43 1 26007 22529
Email: sales.publications@iaea.org • Web site: www.iaea.org/publications